AF614709

Methods in Molecular Biology™

For further volumes:
http://www.springer.com/series/7651

Tiling Arrays

Methods and Protocols

Edited by

Tin-Lap Lee

Reproduction, Endocrinology and Development Program,
School of Biomedical Sciences, The Chinese University of Hong Kong, Shatin,
Hong Kong SAR, China

Alfred Chun Shui Luk

Reproduction, Endocrinology and Development Program,
School of Biomedical Sciences, The Chinese University of Hong Kong, Shatin,
Hong Kong SAR, China

Editors
Tin-Lap Lee
Reproduction, Endocrinology and Development Program
School of Biomedical Sciences
The Chinese University of Hong Kong
Shatin, Hong Kong SAR, China

Alfred Chun Shui Luk
Reproduction, Endocrinology and Development Program
School of Biomedical Sciences
The Chinese University of Hong Kong
Shatin, Hong Kong SAR, China

ISSN 1064-3745 ISSN 1940-6029 (electronic)
ISBN 978-1-62703-606-1 ISBN 978-1-62703-607-8 (eBook)
DOI 10.1007/978-1-62703-607-8
Springer New York Heidelberg Dordrecht London

Library of Congress Control Number: 2013946424

Printed on acid-free paper

Humana Press is a brand of Springer
Springer is part of Springer Science+Business Media (www.springer.com)

Preface

With over 23,000 scientific publications, microarray technology represents an indispensible tool in genome discovery and continues to contribute to novel findings along with other high-throughput technologies. The application of microarrays becomes very popular in academic labs as well as in clinics and commercial entities. During the last decade, the design and coverage of microarrays have gone through a drastic development. Conventional microarrays were largely applied to gene expression study. The density was usually low and the probes were limited to known annotations, which led to incomplete and biased conclusions. The advent of tiling microarrays not only addressed the weakness of conventional microarrays but also provided a new avenue for other genomic applications.

Departing from the biased gene-centric probe design approach, probes on tiling microarrays rely purely on genomic DNA sequence information. The probes are evenly distributed across the nonrepetitive portion of the genome. Commercially available tiling array platforms offer flexible designs with a range of probe lengths and spacing options. According to Gene Expression Omnibus (GEO) at National Center for Biotechnology Information (NCBI), more than 1,100 tiling microarray platforms are now available for various unbiased genomic discoveries.

Typical tiling microarray applications include transcriptome landscape discovery through mapping of the entire transcribed populations, identifying protein binding and methylated regions by chromatin immunoprecipitation (ChIP-chip) and Methylated DNA immunoprecipitation (MeDIP), and identifying genomic origins of replication. Such versatile and robust nature has contributed significantly in decoding the dark matter in the genome, which provides supporting evidence of widespread transcription and regulation events. These include the discovery of novel long noncoding RNAs to the identification of spatiotemporal patterns of gene expression, to the characterization of the transcriptome in various developmental stages and models. Despite high-throughput sequencing methods have recently demonstrated distinct advantages over array-based approaches with regard to data resolution, the "big data" generated requires steep requirements on computer hardware and complex bioinformatics analysis. Tiling microarrays could fill this gap with many established supporting software. Also, the availability of large tiling array reference datasets in major public repositories like GEO, ArrayExpress, and The Encyclopedia of DNA Elements (ENCODE) will allow rapid and straightforward meta-analysis.

This volume focuses on the key applications of tiling microarrays in genomic discovery, including transcriptional regulation, expression, and genetic and epigenetic regulation. Given the applications and types of microarray technology become more numerous and diverse, the limited consensus of how to analyze and interpret the data becomes more apparent. To cope with such challenging issues, the chapters in the data analysis section cover techniques and skills on tiling microarray data analysis that allow readers to leverage the power of dataset from novel perspectives and platform-independent tools. It is our hope that this volume will provide a comprehensive and a down-to-earth approach for maximizing reader's knowledge on this topic.

China, People's Republic

Tin-Lap Lee
Alfred Chun Shui Luk

Contents

Contributors

TAKESHI AKAMA • *Leprosy Research Center, National Institute of Infectious Diseases, Tokyo, Japan*
CHUNG-IL AN • *Division of Cardiovascular Medicine, Department of Internal Medicine, University of California, Davis, Davis, CA, USA*
WAI-YEE CHAN • *Laboratory of Clinical Genomics, Eunice Kennedy Shriver National Institute of Child Health and Human Development, National Institutes of Health, Bethesda, MD, USA*
HOI-HUNG CHEUNG • *Laboratory of Clinical and Developmental Genomics, Eunice Kennedy Shriver National Institute of Child Health and Human Development, National Institutes of Health, Bethesda, MD, USA*
JANEK CLAUS • *Division of Information Technology, Eunice Kennedy Shriver National Institute of Child Health and Human Development, National Institutes of Health, Bethesda, MD, USA*
MARIA COLOMÉ-TATCHÉ • *Faculty of Mathematics and Natural Sciences, Groningen Bioinformatics Centre, University of Groningen, AG Groningen, The Netherlands*
VINCENT COLOT • *Institut de Biologie de l'Ecole Normale Supérieure, Centre National de la Recherche Scientifique (CNRS) UMR8197-Institut National de la Santé et de la Recherche Médicale (INSERM), Paris, France*
DIANA COMAN • *Plant Biotechnology, Department of Biology, ETH Zurich, Zurich, Switzerland*
SUDIPTO DAS • *Cancer Genetics, Department of Molecular and Cellular Therapeutics, Royal College of Surgeons in Ireland, Dublin, Ireland; The National Children's Research Centre, Our Lady's Children's Hospital, Crumlin, Dublin, Ireland*
RUUD DELWEL • *Department of Hematology, Erasmus University Medical Center, Rotterdam, The Netherlands*
XING-WANG DENG • *Department of Molecular, Cellular and Developmental Biology, Yale University, New Haven, CT, USA*
SAINITIN DONAKONDA • *Biotechnologisches Zentrum, Technische Universität, Dresden, Germany*
JOHANNES EICHNER • *Center for Bioinformatics, University of Tuebingen, Tübingen, Germany*
WILHELM GRUISSEM • *Plant Biotechnology, Department of Biology, ETH Zurich, Zurich, Switzerland; Functional Genomics Center Zurich, ETH Zurich, Zurich, Switzerland*
NOBUKO HAGIWARA • *Division of Cardiovascular Medicine, Department of Internal Medicine, University of California, Davis, Davis, CA, USA*
LARS HENNIG • *Plant Biotechnology, Department of Biology, ETH Zurich, Zurich, Switzerland; Functional Genomics Center Zurich, ETH Zurich, Zurich, Switzerland; Department of Plant Biology and Forest Genetics, Uppsala BioCenter, Swedish University of Agricultural Sciences and Linnean Center for Plant Biology, Uppsala, Sweden*
DAVID C. HENSHALL • *Department of Physiology and Medical Physics, The Royal College of Surgeons in Ireland, Dublin, Ireland*

FRANK JOHANNES • *Faculty of Mathematics and Natural Sciences, Groningen Bioinformatics Centre, University of Groningen, AG Groningen, The Netherlands*
YUN-FAI CHRIS LAU • *Laboratory of Cell and Developmental Genetics, Department of Medicine and Institute for Human Genetics, VA Medical Center, University of California, San Francisco, USA*
TIN-LAP LEE • *Reproduction, Endocrinology and Development Program, School of Biomedical Sciences, The Chinese University of Hong Kong, Shatin, Hong Kong SAR, China*
CHRISTOPHE LEMETRE • *Department of Genetics, Albert Einstein College of Medicine, Bronx, NY, USA*
LEI LI • *Department of Biology, University of Virginia, Charlottesville, VA, USA*
YUNMIN LI • *Laboratory of Cell and Developmental Genetics, Department of Medicine and Institute for Human Genetics, VA Medical Center, University of California, San Francisco, CA, USA*
ALFRED CHUN SHUI LUK • *Reproduction, Endocrinology and Development Program, School of Biomedical Sciences, The Chinese University of Hong Kong, Shatin, Hong Kong SAR, China*
HAIYIN LIU • *Faculty of Mathematics and Natural Sciences, Groningen Bioinformatics Centre, University of Groningen, AG Groningen, The Netherlands*
SUZANNE F.C. MILLER-DELANEY • *Department of Physiology and Medical Physics, The Royal College of Surgeons in Ireland, Dublin, Ireland*
KAZUAKI NAKAMURA • *Department of Pharmacology, National Research Institute for Child Health and Development, Tokyo, Japan*
SATYANARAYANA M.R. RAO • *Molecular Biology and Genetics Unit, Chromatin Biology Laboratory, Jawaharlal Nehru Centre for Advanced Scientific Research, Bangalore, India*
OWEN M. RENNERT • *School of Biomedical Sciences, The Chinese University of Hong Kong, Shatin, Hong Kong SAR, China*
CHANDAN SASTRY • *Division of Information Technology, Eunice Kennedy Shriver National Institute of Child Health and Human Development, National Institutes of Health, Bethesda, MD, USA*
SUMEETA SINGH • *Division of Information Technology, Eunice Kennedy Shriver National Institute of Child Health and Human Development, National Institutes of Health, Bethesda, MD, USA*
SWATI SINHA • *Molecular Biology and Genetics Unit, Chromatin Biology Laboratory, Jawaharlal Nehru Centre for Advanced Scientific Research, Bangalore, India*
RAYMOND L. STALLINGS • *Cancer Genetics, Department of Molecular and Cellular Therapeutics, Royal College of Surgeons in Ireland, Dublin, Ireland; The National Children's Research Centre, Our Lady's Children's Hospital, Crumlin, Dublin, Ireland*
KOICHI SUZUKI • *Leprosy Research Center, National Institute of Infectious Diseases, Tokyo, Japan*
AKITO TANOUE • *Department of Pharmacology, National Research Institute for Child Health and Development, Tokyo, Japan*
ERDOGAN TASKESEN • *Department of Hematology, Erasmus University Medical Center, Rotterdam, The Netherlands*

RENÉ WARDENAAR • *Faculty of Mathematics and Natural Sciences, Groningen Bioinformatics Centre, University of Groningen, AG Groningen, The Netherlands*

BAS WOUTERS • *Department of Hematology, Erasmus University Medical Center, Rotterdam, The Netherlands*

HUIYONG ZHANG • *Department of Biology, University of Virginia, Charlottesville, VA, USA*

ZHENGDONG D. ZHANG • *Department of Genetics, Albert Einstein College of Medicine, Bronx, NY, USA*

MING ZHENG • *Department of Anesthesia, Stanford University School of Medicine, Stanford, CA, USA*

LIHUA JULIE ZHU • *Program in Gene Function and Expression; Program in Bioinformatics and Integrated Biology; Program in Molecular Medicine, University of Massachusetts – Medical School, Worcester, MA, USA*

Part I

Overview

Chapter 1

A Brief Introduction to Tiling Microarrays: Principles, Concepts, and Applications

Christophe Lemetre and Zhengdong D. Zhang

Abstract

Technological achievements have always contributed to the advancement of biomedical research. It has never been more so than in recent times, when the development and application of innovative cutting-edge technologies have transformed biology into a data-rich quantitative science. This stunning revolution in biology primarily ensued from the emergence of microarrays over two decades ago. The completion of whole-genome sequencing projects and the advance in microarray manufacturing technologies enabled the development of tiling microarrays, which gave unprecedented genomic coverage. Since their first description, several types of application of tiling arrays have emerged, each aiming to tackle a different biological problem. Although numerous algorithms have already been developed to analyze microarray data, new method development is still needed not only for better performance but also for integration of available microarray data sets, which without doubt constitute one of the largest collections of biological data ever generated. In this chapter we first introduce the principles behind the emergence and the development of tiling microarrays, and then discuss with some examples how they are used to investigate different biological problems.

Key words Tiling microarrays, ChIP-chip, Transcriptome mapping, Gene expression, Probe design

1 Introduction

Large-scale analysis of genomic data ushered in a new era of biomedical research. With the ability to investigate whole genomes, it is now possible to evaluate and infer the degree of expression of a large number of genes, or even entire genomes. The latest technological achievements in genomics, biotechnology, informatics, and miniaturization enabled the emergence of the microarray technology. So far, traditional microarrays were mainly used to identify transcripts from samples in a specific context or condition. It then became possible to monitor the level of expression of some particular genes. Knowing the relative abundance of a transcript can indeed reveal the degree of expression of its gene. Their application

Tin-Lap Lee and Alfred Chun Shui Luk (eds.), *Tiling Arrays: Methods and Protocols*, Methods in Molecular Biology, vol. 1067,
DOI 10.1007/978-1-62703-607-8_1, © Springer Science+Business Media New York 2013

allowed as well a high-throughput comparison between samples, aiming to identify discrepancy in patterns of expression between different conditions and ultimately identify biomarkers for potential diagnostic or therapeutic purposes.

The microarray technology has its root and founding principles in the Southern blotting technique developed by Edwin Southern in 1975 [1], but it provides a massive scale-up in the nucleic acid hybridization assay because of its huge capacity resulting from miniaturization and automation [2]. The development of cDNA libraries [3] made it possible to spot cDNA probes onto a solid surface inside microscopic wells, allowing the high-throughput screening of thousands of genes simultaneously. In one of the very first microarray studies, Schena et al. in 1995 used spotted DNA microarrays to measure the differential expression of 45 genes in *Arabidopsis thaliana* [3]. Despite such an impressive scale-up, the early DNA microarrays had an obvious technological limitation: because the probes need to be designed to target specific genes [4], the annotation of which may be incomplete and erroneous [2], incorporating annotation data into the experimental design could add errors and bias. As a result, an ideal microarray technology for genome-wide screen should avoid any prior genome or gene annotations.

Tiling microarrays share the same biophysical principle with traditional genomic microarrays by hybridizing labeled fragments of cDNA or RNA to small polymers of nucleotides (probes) attached to solid glass surfaces. However, they use an ***unbiased*** approach: unlike traditional microarrays, they preclude a priori annotation information. In addition, the probes used in tiling microarrays can overlap, sometimes with only several nucleotide-shifts (e.g., single-base offset) [5, 6]. They also comprehensively represent large portions of genome, or even entire genomes in the case of high-density tiling microarrays (Fig. 1). Instead of probing for known or predicted gene sequences, tiling microarrays map genome sequences that exist contiguously in the genome, or portion of the genome like chromosomes, in a totally unbiased fashion. Today, combined with complete whole-genome sequences, the DNA microarray technology can be used in novel experimental approaches to investigate different biological problems on the whole-genome level [2].

In this chapter, we give a general introduction to tiling microarrays, including their different types and probe designs. We also present their most popular forms of applications with examples and discuss their limitations. The data analysis step is also explored, discussing some of the most common tools employed to analyze the final output from tiling array experiments.

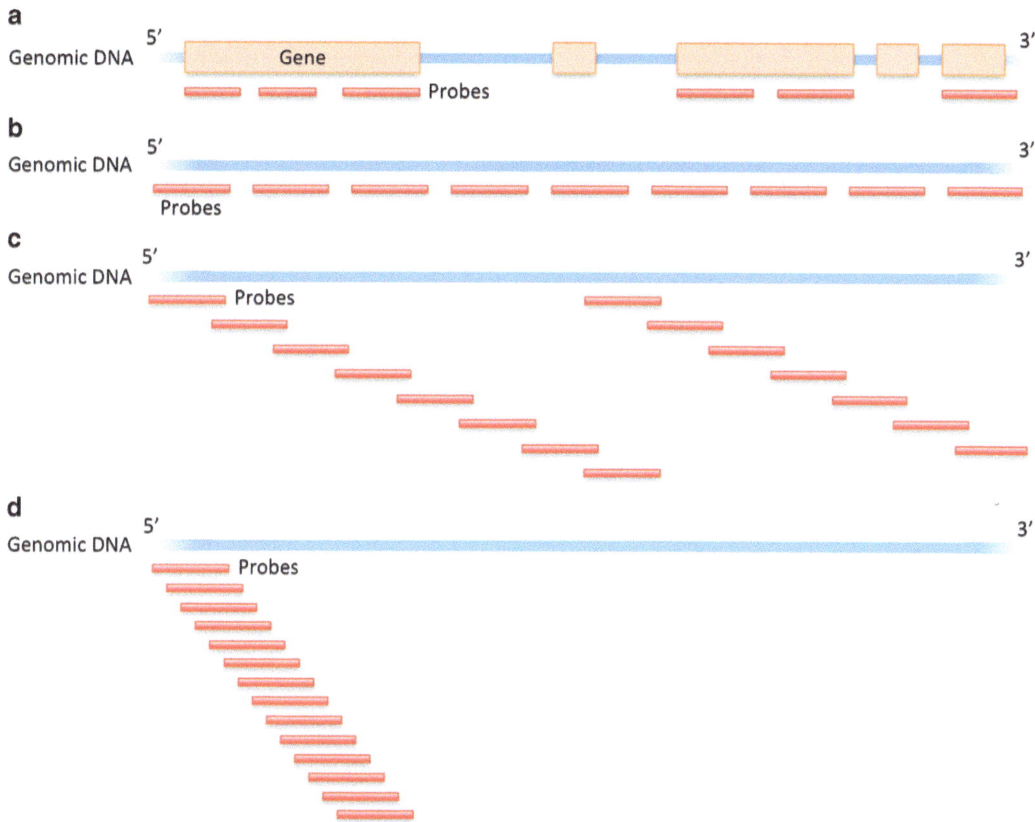

Fig. 1 Different probe designs and tiling paths. (**a**) Classical gene expression microarray. (**b**) Tiling microarray with nonoverlapping probes leaving few base-pair gaps between probes. (**c**) Tiling microarray with overlapping probes. The overlaps are several nucleotides long. (**d**) Tiling microarray with 1-nucleotide overlapping probes, which offers the highest resolution

2 Probe Design and Manufacturing of Tiling Microarrays

Like the original genomic microarrays, tiling arrays rely on the principle of hybridization of labeled molecules of interest to probes attached to a solid surface. This principle, first described in the seminal paper by Watson and Crick in 1953 [7, 8], is the heteroduplex formed by complementary base pairing. However, tiling arrays consist of a more comprehensive and unbiased presentation of the genome than gene microarray. With the constantly increasing number of available complete genome sequences, it is now possible to design contiguous probes without relying on any potentially erroneous annotation [9]. Additionally, the organization of the probes can be arranged for higher or lower resolution depending on the gap and span of overlap between the probes to suit the need of a particular experiment.

2.1 Probe Design

Designing probes to represent a complete or partial genome is the first step in a high-density tiling arrays experiment. It poses a challenge to computational biologists, as they need to consider several aspects of the microarray experiment and answer the following three questions [10–12]:

- How to avoid nonspecific hybridization of molecules to the probes?
- How to maximize hybridization with the biophysical properties of the probe sequences?
- Which probe design is best for the expected outcome of the tiling microarray experiment?

2.1.1 Specificity and Uniqueness of Probes

Due to its nonrandom nature, a genome contains many redundant sequences [10]. Consequently, cross-hybridization between probes and nonspecific DNA/RNA molecules can become a serious problem for a hybridization-based method, as it can result in significant background noise [12]. This problem is especially grave with a genome-wide coverage and becomes even more critical with shorter probe sequences [10]. Alignment-based tools (with BLAST or BLAST-like algorithms [13, 14]) can provide a solution to overcome this problem where potential probes are tested for their specificity by alignment algorithm against a sequence database [10, 15]. RepeatMasker is the most widely used algorithm for repeat detection, identifying repeat regions in genomes using a database of repeat elements [16]. Repetitive sequences can then be masked before probe design.

2.1.2 Thermodynamic and Biophysical Properties of the Probes

An important determinant of how probes behave towards hybridization is their thermodynamic property. This property determines the probe melting temperature and thus, in turn, affects the hybridization process [17]. The probe design also needs to take into consideration the possibility of probe self-hybridization [12]. Another aspect to be considered is the potentially damaging biochemical properties that are required by the manufacturing process [18]. Given the sequence of a particular probe, its thermodynamic characteristics can be determined by well-characterized formulas taking into account either the nucleotide composition of the sequence or the exact sequence in itself [10].

2.1.3 Experimental Design

The final aim of the overall experiment is also critical in the choice of the platform and the design of the probes. A transcriptional mapping study will be more efficient using overlapping tiles and high density [19], whereas copy number variation studies may benefit from a sparse tile path [20]. ChIP-chip methods have also been described to be more efficient with high-density tiling microarrays with consistent spacing of probes, particularly for DNA–protein-binding site identification [21].

Several software solutions have been developed to provide efficient tools for designing high-density tiling array probes with unique sequences. Some solutions have been put forward as integrative platforms in order to optimize the design of probes. For example, Gräf et al. described a method defining a uniqueness score with the content of the shortest substring of the probe and its similarity to any other portion in the genome [12]. The probes designed with this method were subsequently validated with BLAT and further selected with their thermodynamic features taken into account. An alternative presented by Bertone et al. used dynamic programming to infer tile paths and determine most optimal probe sequence using a heuristic approach to reduce dataspace complexity [10]. Recently, ChipD [11] was developed as a Web server-based solution for probe design, providing the user with the sequence of probes, their genomic locations, as well as their hybridization features. However, there is still potential for further improvements in the probe design step of tiling microarray experiments.

2.2 Manufacturing

Depending on their design and manufacturing, tiling microarrays can be classified into two types. The first type of tiling microarrays was manufactured by spotting PCR products as probes on the solid surface, a method also used to make traditional microarrays. This type of tiling microarrays was developed and used in the first half of the 2000s [12]. In some studies, they were successfully used to identify DNA–protein interactions [22, 23] or histone modifications [24] in certain part of the human genome. Despite such successes, the limitations of this type of tiling microarray in genomic coverage and resolution are evident and led to the emergence of a second type. High-density oligonucleotide microarrays consist of short probes (about 100 nucleotides) that are directly synthesized in situ onto the solid surface of the microarray. This synthesis can be performed using a photolithographic mask [25–27], but other techniques exist, such as ink-jet [28]. This type of array can present several millions of spots, each containing several millions of the same probe. This second type has been shown to outperform PCR-product arrays [24].

Several companies, as leaders in the microarray technology, have developed DNA chips for tiling microarray experiments. Currently, Affymetrix has issued a tiling microarray with six million 25-mer probes, offering very high definition. They propose three different platforms for human. The GeneChip Human tiling 1.0 Array Set consists of 14 arrays, designed to cover the entire human genome, devoted for transcript mapping experiment. The GeneChip Human tiling 2.0 Array Set, with 7 arrays, has been designed for ChIP-chip application; and the GeneChipHuman Promoter 1.0 consists of a subset of probes of the previous microarray for promoter regions. These tiling arrays are designed for an optimal specific hybridization, using RepeatMasker [29], containing over 6.5 million probes per set, and with 25-mer oligonucleotides with about 10-base pair gap between tiles, leading to a 35-base pair resolution [30]. Roche/NimbleGen have also developed their tiling platforms, which usually have lower probe densities than Affymetrix products. For comparative genomic hybridization, they offer nine different platforms with up to 4.2 million 50- to 75-mer probes and 284-bp gaps between probes, offering above 1.4-kbp resolution for CNV discovery. For ChIP-chip experiments, they offer a 10-array set and a 4-array set with 2.1 million probes. They both spot 50- to 75-mer probes with a median gap between probes of 100 bp for the first one and 205 bp for the second. They also offer whole-genome tiling arrays in a similar format for DNA-methylation studies [30]. Similar formats and settings are also available from Agilent, but they usually present lower density. For example, their Human ENCODE ChIP-chip microarray presents over 153,000 probes, and their CGH bundles offer up to 1.1 million probes [30].

3 Applications

3.1 Protein–DNA Interaction Profiling

Gene expression is highly regulated by many mechanisms, particularly via DNA-binding of transcription factors and other molecules. Chromatin immunoprecipitation arrays (ChIP-chip) represent the most popular form of application of tiling microarrays. This particular technique, which couples chromatin immunoprecipitation (ChIP) and microarray techniques (or "chip"), aims to screen on a genome-wide scale the binding sites of transcription factors and other DNA-binding molecules and proteins [31]. Through its determination of transcription factor-binding sites, ChIP-chip can shed new light on regulatory patterns of gene expression [32] and thus provide extremely important information towards a better understanding of critical regulation and gene functions [12].

A ChIP-chip experiment involves cross-linking proteins and their bound DNA by formaldehyde, followed by immune-precipitation, and shearing DNA into ~500-bp fragments (Fig. 2a) [33]. The protein of interest (e.g., a transcription factor), bound to its DNA-binding sites, will precipitate with its bound DNA fragments in the presence of its antibody. The fragments of DNA co-precipitated with the protein of interest are then labeled (e.g., using a fluorescent tag) and hybridized to a microarray. The image is processed to detect the signal of the labeled fragments hybridized on the microarray [34]. The probes with ChIP signals can also be mapped back to the genome [10] in order to identify genome-wide protein-binding sites. Due to the nature of this approach, a major limitation is the quality of the antibody and its affinity to the protein of interest. Indeed, the quality of the result is highly dependent on the quality of the protein precipitation by the antibody [32].

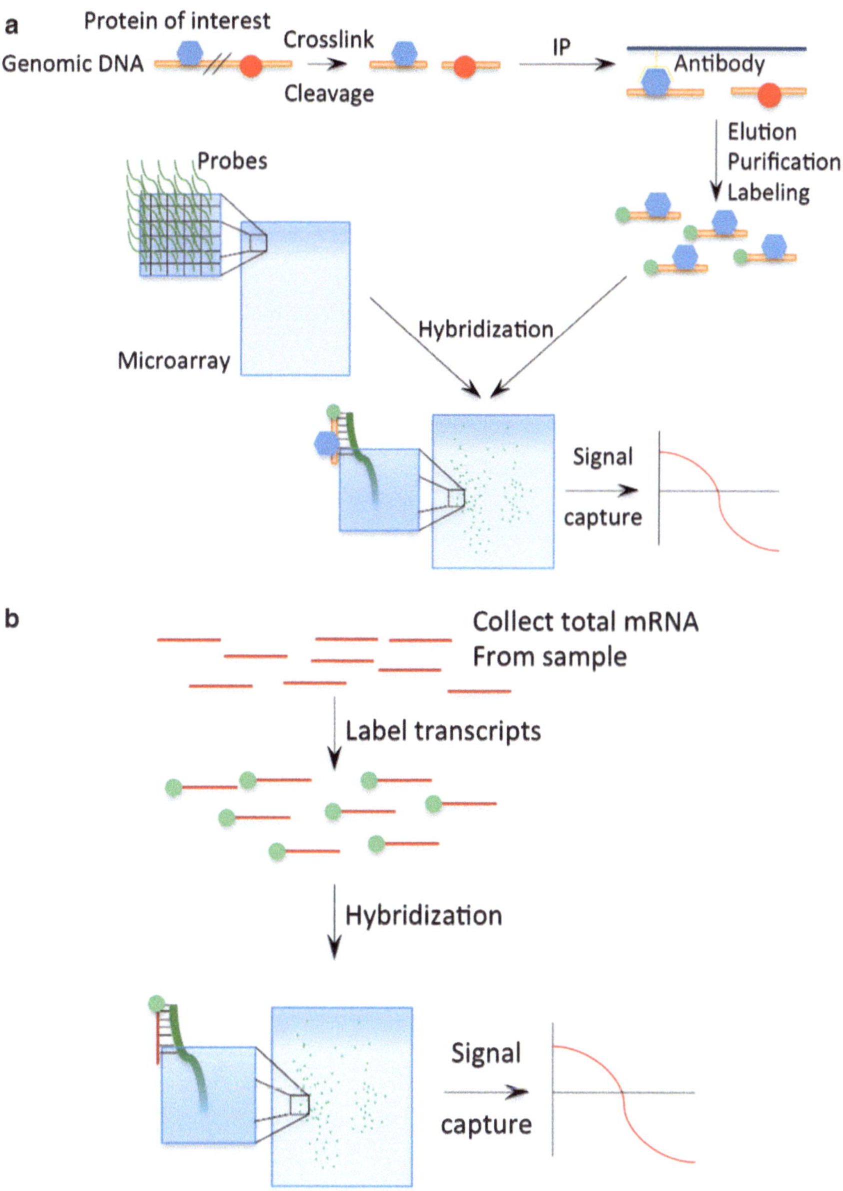

Fig. 2 Schematic workflow summarizing different tiling microarray applications. (**a**) ChIP-chip application of tiling arrays; after cross-link of DNA with proteins, the strands are cleaved and the fragments presented to the antibody specific to the protein of interest (in *blue*). The antibody retains only the fragments with the protein of interest and the others are eluted. The retained fragments linked to the protein are then hybridized to the tiling microarray and mapped to the genome. (**b**) A transcriptome mapping application of tiling microarray where mRNAs from the sample are extracted and labeled and then hybridized to the tiling microarray. (**c**) In nucleosome localization, the internucleosomal DNA is firstly digested with DNaseI, leaving only nucleosomes attached to their DNA. The octamer of protein is then eliminated to only leave out the nucleosomal DNA fragments that are then labeled. The fragments are then hybridized to the microarray and mapped back to the genome. (**d**) An arrayCGH experiment firstly isolates genomic DNA from a patient and from a control, which are then differentially labeled. The two sets of labeled DNA are then presented to the tilling microarray for competitive hybridization. The signal obtained compares the different copy numbers between the two sets of samples and deciphers the copy number variations

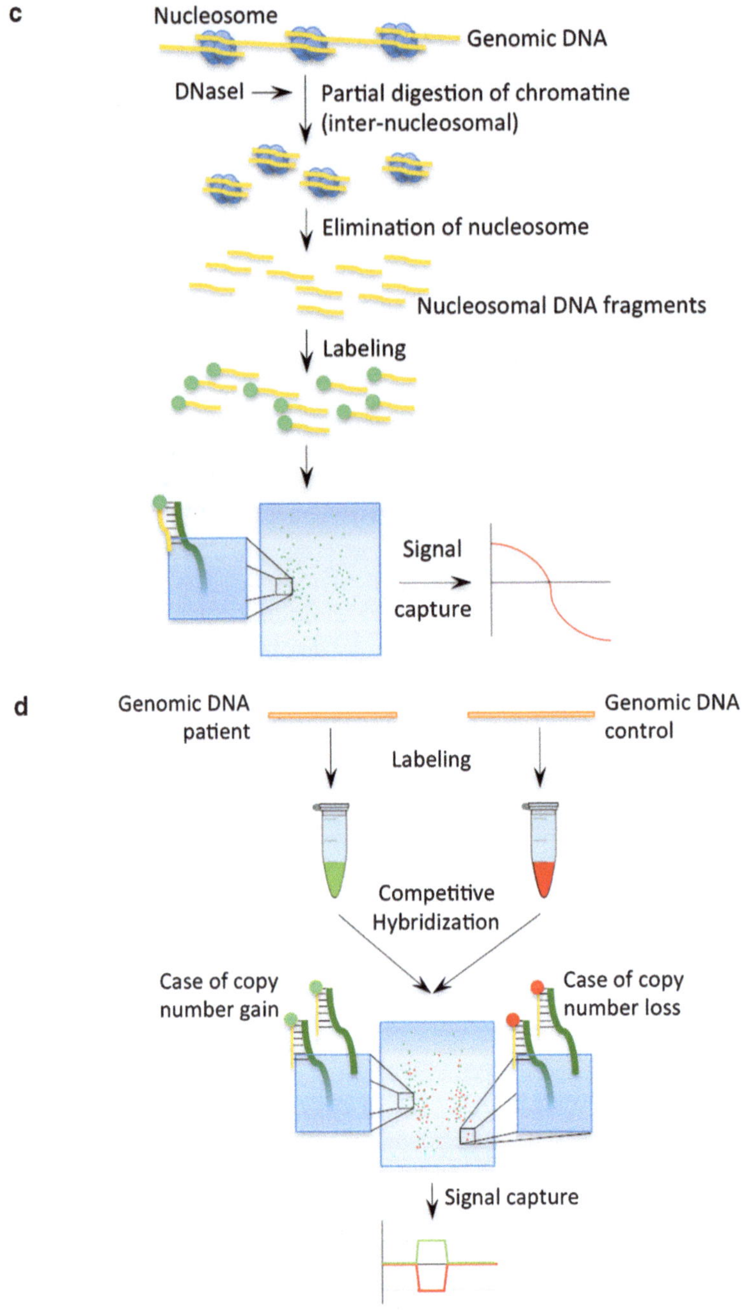

Fig. 2 (continued)

Multiple studies used ChIP-chip approach to identify proteins and transcription factor DNA-binding sites. Amongst the first applications of ChIP-chip were studies screening the genomic binding sites of 16 transcription factors in yeast [35, 36]. In a recent review [37], Peggy Farnham emphasized the great insights brought by genomic studies using ChIP-chip (and lately ChIP-seq), noting the confirmation that RNA polymerase II and transcription factors bind almost exclusively to proximal promoters that are weakly active, and also that the binding of a transcription factor to enhancer regions may not be sufficient for regulation, hence promoting the idea of a "cooperation" of regulators.

3.2 Gene Expression and Transcriptome Mapping

Gene expression mapping is the other most common application of tiling microarrays. The underlying idea is to hybridize reverse-transcribed RNA into cDNA to tiling microarrays in order to identify the transcribed portions of the genome. Because using tiling microarrays does not involve prior gene annotation, it would be an alternative approach towards gene prediction in a systematic fashion. Traditional in silico gene prediction approaches are known to be error-prone and can miss some rare or condition-specific transcripts [25]. The high resolution of tiling microarrays can overcome some of these limitations. For instance, their sensitivity makes it possible to identify rare transcripts. Also, thanks to their overlapping probes, it is possible to obtain a much more accurate picture of the predicted gene structure.

For this approach, RNAs are extracted, purified (with rRNA depletion), and reverse transcribed into cDNA (Fig. 2b). The cDNA molecules are then labeled and hybridized to a tiling microarray. The transcripts observed can then highlight the portion of the genome from which they originate, therefore providing useful information about the gene position or structure.

As stressed in a review by Mockler & Ecker [25], the power and resolution of this approach for gene and gene structure prediction can be observed with some of the first applications of high-density oligonucleotide tiling arrays to transcription mapping on human chromosomes 21 and 22 [38–40]. They used 25-mer probes with a gap of 35 bp between probes across the entire chromosomes 21 and 22. They were able to demonstrate that it was possible to identify up to ten times more transcripts than what was previously thought and predicted, and that about 90 % of the observed transcription was taking place in non-previously annotated regions. In another early application of tiling microarrays to transcriptome mapping [41], Kapranov et al. identified new classes of small RNAs (sRNAs) using a high-resolution tiling microarray (5-nucleotide resolution). Two of these new types of sRNAs were found to cluster at each extremity of the gene. A third class of sRNAs, which skip most of the exons, was also identified at the 5′ limit of protein-coding genes.

In a recent study, Spencer et al. used tiling microarrays to examine the spatiotemporal motifs of expression of genes in *Caenorhabditis elegans* [42]. Using tiling arrays to profile gene expression in more than 30 cells at different developmental stages of the organism, they discovered novel transcription regions in noncoding domains that count for about 10 % of the genome. They also showed a differential pattern of expression of about 75 % of the transcripts between the different developmental stages. A similar approach has also been applied to *Drosophila melanogaster* [43].

High-density tiling microarrays can also be used to identify alternatively spliced forms of a transcript. Like DNA-methylation, alternative splicing plays an important role in genetic regulation. It allows a single gene to code multiple functionally distinguished proteins with different amino acid sequences [44]. Moreover, this expansion of the expression potential of a gene is further increased with the different types of alternative splicing that have been observed, including exon skipping or intron retention [32]. Tiling microarrays offer the potential for screening alternative splicing and discover de novo transcript isoforms. In theory, tiling microarrays can identify any type of alternative splicing. However, the probe density can limit the sensitivity of the detection of some alternative splicing events [44]. The unbiased coverage of whole genome offered by tiling microarrays enables the detection of different exons present in the transcripts. Hybridization signals from different exons are useful to elucidate gene structures, and subsequent comparison between signals can help to identify the different isoforms existing in the sample. Recently Eichner et al. presented a novel computational approach based on support vector machines to identify intron retention and exon skipping [44].

3.3 Identification of Genomic Copy Number Variations

Microarray-based comparative genome hybridization (array-CGH) technique uses microarray to compare the genome from the control sample with the disease sample to identify the genomic segments with copy number variations (CNVs). This approach is often chosen for diagnostic purposes, since comparison between two types of samples is possible (e.g., cancer cells versus control cells). The differential hybridization can therefore highlight a potential biomarker of interest that presents a different pattern between the two sets of samples. The methodology is straightforward, consisting of labeling the control and test samples with different fluorescence dyes and hybridizing them to the tiling microarray. The fluorescence is then captured and processed. The difference between fluorescence from the sets of samples is calculated and allows inferring the CNV between the samples (Fig. 2c).

Sebat et al. were amongst the first to report the use of tiling microarrays to identify differences in copy number of genome portions between samples from different tissues and lymphoblastoid cell lines [45]. They observed different patterns between the samples of the same individual and found somatic mutations in gene clusters of T cell receptors and immunoglobulins. In a recent study, Brommesson et al. used array-CGH and hierarchical clustering to show that one out of five pairs of unilateral breast carcinoma shared copy number profile while only one of eight pairs of bilateral tumors showed similar genomic features [46]. This result shed new light on the genomic origin of unilateral and bilateral primary breast tumors. In their review, Shinawi and Cheung discussed the important applications of array-CGH, stressing the tremendous improvements in diagnosis for numerous conditions including autism and dysmorphism [47]. The possibility to measure the relative difference of abundance of CNVs between two samples has enhanced the potential to discover new patterns and new biomarkers.

3.4 Epigenomics

3.4.1 MeDIP-Chip

As a major process of epigenetics, DNA methylation has been intensively studied to establish its involvement in gene expression regulation. With the advent of tiling microarrays, it has been possible to map DNA methylation sites on a genome-wide scale, which is the principle of MeDIP-chip. Similar to ChIP-chip, MeDIP-chip uses immunoprecipitation with an anti-5-methyl-cytosine antibody to specifically precipitate the genomic segments containing methylated DNA and tiling microarray to hybridize the precipitated DNA.

Studies of DNA methylation with tiling microarrays provided the first so-called *methylomes* and advanced our understanding of how gene expression is affected by epigenetic modifications and regulations. First achieved in *Arabidopsis* [48, 49], these pioneer experiments in methylome discovery showed that highly repetitive DNA sequences presented the highest density of DNA methylation. In a recent study, Rauch et al. used tiling microarrays to examine DNA methylation in human B cells [50] and found that most of the methylated regions were associated with genes.

3.4.2 ChIP-Chip

ChIP-chip experiments can also provide insights into epigenomic components. As demonstrated by the literature, studies of histone modification remain one of the main applications of ChIP-chip. In 2005, Berstein et al. developed a comparative analysis of histone modification between human and mouse using a tiling microarray-based ChIP-chip approach [51].

3.4.3 Nucleosome Localization

Nucleosomes are the basic units in the process of chromatin compaction in eukaryotes, in which DNA strand is wrapped around histone octamers. Tiling microarrays can also be used to study nucleosome localization, which is essential for a better understanding of how DNA-binding proteins such as transcription factors get access to their binding sites on the DNA. This technique involves partial digestion of the chromatin in order to eliminate the linker DNA region between two nucleosomes. After removal of the proteins, the nucleosomal DNA is isolated, labeled, and subsequently hybridized to the tiling microarray (Fig. 2d). This approach requires a high sensitivity and resolution for the tiling microarray due to the size of the portion of nucleosomal DNA (146 or 147 nucleotides) [33].

4 Processing and Analysis of Tiling Microarray Data

The large amount of data generated by a tiling microarray experiment, the complexity, and the framework of the study necessitate applying specific algorithms to extract the relevant information expected from the experiment. Even though many software packages have been developed to analyze complex data sets generated by high-density tiling microarrays, there still remains much room for improvement and innovation in algorithms and methods of tiling microarray data analysis. And indeed, novel approaches still appear in the literature.

Amongst the most popular software tools for tiling microarray data analysis is Tiling Analysis Software (TAS), developed by Affymetrix for its own microarray platforms [52]. With TAS, the differential expression between the probe intensities within a given window is estimated with a Wilcoxon rank test. Kampa et al. also implemented a probe local expression level estimation method in TAS [40]. An alternative to TAS as a peak-detection algorithm is MAT (Model-Based analysis of Tiling arrays) [53]. MAT normalizes the signal from probes using a Mixture Model. It is commonly applied to ChIP-chip data to identify DNA-binding sites, however, other groups have reported their successful adaptation of MAT to transcriptome mapping data [54, 55]. MAT first estimates a baseline of the signal from the probes using a basic linear modeling approach considering two parameters: features from the probe sequences and the copy numbers. This baseline allows pre-filtering most of the noise and reveals the actual significant signal from the probes. Subsequently, MAT standardizes each probe on the microarray using a probe behavior model and the estimate of the baseline, providing a *t*-value representing the actual behavior of the probe from its predicted one and therefore stressing a potential ChIP-enriched portion. TileProbe integrates publicly available data to maximize the outcome from MAT [56]. Besides applications specifically designed for Affymetrix platforms, alternative methods have been developed for NimbleGen microarray platforms. TAMAL has been described as the algorithm of choice for the detection of the binding sites in a ChIP-chip experiment for these platforms [57].

TileScope, an online pipeline for tiling microarray data analysis, is an example of software with a user-friendly interface [58]. It can normalize signals channel-wise and between arrays, calculate a score for each probe, and identify genomic regions with significant signal-to-noise ratios. Starr, another fully integrated pipeline, can be used for exploration, visualization, or quality control [59]. After peak detection in a ChIP-chip experiment, motifs in the transcription factor-binding sites can be identified by MEME [33, 60].

TileShuffle has been recently developed to identify noncoding RNAs in tiling microarray experiments [52]. The algorithm identifies statistically significant, differentially transcribed, and differentially expressed portions from a background distribution. Many statistical algorithms are available for tiling microarray data analysis to identify alternative splicing. MIDAS [61], the most popular method, uses an ANOVA test to distinguish signal from alternatively spliced exons. This statistic test is based on the assumption that the signal intensity of an exon is relatively constant compared to the signal intensity of the overall gene. As mentioned earlier, methods based on machine learning have also been recently developed to analyze tiling microarray data, particularly for alternative splicing detection. For example, Eichner et al. used the support vector machine (SVM), a supervised machine learning technique, to analyze alternative splicing data from tiling microarrays [44].

5 Discussion and Future Directions

Tiling microarrays have been great achievements in biotechnology: for the first time, cells can be queried and analyzed on a whole-genome level and in an systematic fashion. By using no prior annotation data, tiling microarrays offer an full evaluation and screening of protein–DNA interaction, gene expression or structure, and DNA methylation. As a result, they offer a wide panel of potential application from epigenetics to gene expression and structural genetics. They have been successfully applied in these different aspects and lead to important discoveries. Many software tools have been developed for tiling microarray data analysis, each suitable for a specific microarray platform and its application. These software tools facilitate the application of tiling microarrays to address a wide range of different biological questions, whether it is to identify a protein–DNA interaction pattern within the genome or a differential gene expression between samples.

Because tiling microarrays share similar basic principles with traditional genomic microarrays, they also share some of their limitations: both types of microarrays suffer from nonspecific hybridization [52] and narrow signal dynamic range. Intensive normalization is then required to offset both cross-hybridization and background noise [62]. Despite the great promises from the technology of tiling arrays, a literature review of the field does not indicate extensive applications of tiling arrays between 2007 and 2011. This could be explained by the emergence of the next-generation sequencing technologies [63] as shown in Fig. 3. It has also been shown that high-throughput sequencing approaches have clear advantages over tiling array methods [64, 65]. However, tiling

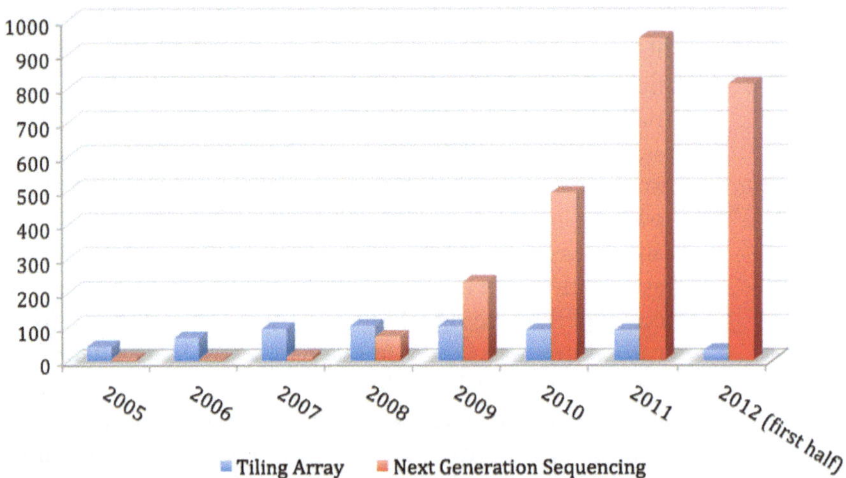

Fig. 3 Number of recent publications related to tiling microarray (*blue bars*) and next-generation sequencing (*red bars*) included PubMed

arrays can still provide an important and meaningful impact in biomedical research, particularly with the development of references and complete genome sequences, and since they have shown to be performing similarly to high-throughput sequencing [63]. In a study in 2009, Sasidharan et al. conducted a comparison between tiling arrays and high-throughput sequencing [63]. This study tried to compare the transcripts identified with tiling arrays and sequencing results, and showed that a significant overlap of identified transcripts exists between the two approaches. But in 2011, Ho et al. published a systemic comparison of a hybridization-based against a sequencing-based ChIP experimental setting (70), in which they showed that ChIP-seq platforms produce better signal-to-noise ratio and better peak-detection with sharper peaks.

From these results and literature, it appears that tiling microarrays have reached a critical step in their evolution, and in the light of alternative innovative approaches, they offer limited advantages. However, they still remain a relatively low-cost option compared to sequencing-based techniques and therefore remain a viable choice for researchers (56), especially when considering they can perform with similar robustness (63).

Acknowledgements

This work was supported by an NIH grant (5R00LM009770-06) from the National Library of Medicine to Z.D.Z.

References

1. Southern EM (1975) Detection of specific sequences among DNA fragments separated by gel electrophoresis. J Mol Biol 98:503–517
2. Lander ES (1999) Array of hope. Nat Genet 21:3–4
3. Schena M et al (1995) Quantitative monitoring of gene expression patterns with a complementary DNA microarray. Science 270: 467–470
4. Licatalosi DD, Darnell RB (2010) RNA processing and its regulation: global insights into biological networks. Nat Rev Genet 11: 75–87
5. Kapranov P, Sementchenko VI, Gingeras TR (2003) Beyond expression profiling: next generation uses of high density oligonucleotide arrays. Brief Funct Genomic Proteomic 2:47–56
6. Jacquier A (2009) The complex eukaryotic transcriptome: unexpected pervasive transcription and novel small RNAs. Nat Rev Genet 10:833–844
7. Watson JD, Crick FH (1953) The structure of DNA. Cold Spring Harb Symp Quant Biol 18:123–131
8. Watson JD, Crick FH (1953) Molecular structure of nucleic acids; a structure for deoxyribose nucleic acid. Nature 171:737–738
9. Yazaki J, Gregory BD, Ecker JR (2007) Mapping the genome landscape using tiling array technology. Curr Opin Plant Biol 10:534–542
10. Bertone P, Trifonov V, Rozowsky JS, Schibert F, Emanuelsson O, Karro J et al (2006) Design optimization methods for genomic DNA tiling arrays. Genome Res 16:271–281
11. Dufour YS, Wesenberg GE, Triltt AJ, Glasner JD, Perna NT, Mitchell JC, Donohue TJ et al (2010) chipD: a web tool to design oligonucleotide probes for high-density tiling arrays. Nucleic Acids Res 38:W321–W325
12. Gräf S, Nielsen FG, Kurtz S, Huynen MA, Birney E, Stunnenberg H, Flicek P (2007) Optimized design and assessment of whole genome tiling arrays. Bioinformatics 23: i195–i204
13. Mount DW (2007) Using the basic local alignment search tool (BLAST). CSH protocols 2007, pdb top17, See http://www.ncbi.nlm.nih.gov/pubmed/21357135
14. Altschul SF, Gish W, Miller W, Myers EW, Lipman DH (1990) Basic local alignment search tool. J Mol Biol 215:403–410
15. Wang X, Seed B (2003) Selection of oligonucleotide probes for protein coding sequences. Bioinformatics 19:796–802
16. Jurka J, Kapitonov VV, Pavlicek A, Klonowski P, Kohany O, Walichieme J (2005) Repbase update, a database of eukaryotic repetitive elements. Cytogenet Genome Res 110:462–467
17. Gresham D, Curry B, Ward A, Gordon DB, Brizuela L, Kruglyak L, Botstein D (2010) Optimized detection of sequence variation in heterozygous genomes using DNA microarrays with isothermal-melting probes. Proc Natl Acad Sci USA 107:1482–1487
18. Lockhart DJ, Dong H, Byrne MC, Follettie MT, Gallo MV, Chee MS et al (1996) Expression monitoring by hybridization to high-density oligonucleotide arrays. Nat Biotechnol 14: 1675–1680
19. Huber W, Toedling J, Steinmetz LM (2006) Transcript mapping with high-density oligonucleotide tiling arrays. Bioinformatics 22:1963–1970
20. Graubert TA, Cahan P, Edwin D, Selzer RR, Richmond TA, Eis PS et al (2007) A high-resolution map of segmental DNA copy number variation in the mouse genome. PLoS Genet 3:e3
21. Qi Y, Rolfe A, Maclsaac KD, Gerber GK, Pokholok D, Zeitlinger J et al (2006) High-resolution computational models of genome binding events. Nat Biotechnol 24:963–970
22. Rada-Iglesias A, Wallerman O, Koch C, Ameur A, Enroth S, Clelland G et al (2005) Binding sites for metabolic disease related transcription factors inferred at base pair resolution by chromatin immunoprecipitation and genomic microarrays. Hum Mol Genet 14:3435–3447
23. Kim TH, Barrera LO, Zheng M, Qu C, Singer MA, Richmond TA et al (2005) A high-resolution map of active promoters in the human genome. Nature 436:876–880
24. Emanuelsson O, Nagalakshmi U, Zheng D, Rozowsky JS, Urban AE, Du J et al (2007) Assessing the performance of different high-density tiling microarray strategies for mapping transcribed regions of the human genome. Genome Res 17:886–897
25. Mockler TC, Chan S, Sundaresan A, Chen H, Jacobsen SE, Ecker JR (2005) Applications of DNA tiling arrays for whole-genome analysis. Genomics 85:1–15
26. Southern EM, Maskos U, Elder JK (1992) Analyzing and comparing nucleic acid sequences by hybridization to arrays of oligonucleotides: evaluation using experimental models. Genomics 13:1008–1017
27. Pease AC, Solas D, Sullivan EJ, Cronin MT, Holmes CP, Fodor SP (1994) Light-generated oligonucleotide arrays for rapid DNA sequence

analysis. Proc Natl Acad Sci USA 91: 5022–5026

28. Hughes TR, Mao M, Jones AR, Burchard J, Marton MJ, Shannon KW et al (2001) Expression profiling using microarrays fabricated by an ink-jet oligonucleotide synthesizer. Nat Biotechnol 19:342–347
29. Tempel S (2012) Using and understanding RepeatMasker. Methods Mol Biol 859:29–51
30. http://www.home.agilent.com
31. Horak CE, Snyder M (2002) ChIP-chip: a genomic approach for identifying transcription factor binding sites. Methods Enzymol 350: 469–483
32. Gregory BD, Yazaki J, Ecker JR (2008) Utilizing tiling microarrays for whole-genome analysis in plants. Plant J 53:636–644
33. Liu XS (2007) Getting started in tiling microarray analysis. PLoS Comput Biol 3:1842–1844
34. Bernstein BE, Humphrey EL, Liu CL, Schreiber SL (2004) The use of chromatin immunoprecipitation assays in genome-wide analyses of histone modifications. Methods Enzymol 376:349–360
35. Lee TI, Rinaldi NJ, Robert F, Odom DT, Bar-Joseph Z, Gerber GK et al (2002) Transcriptional regulatory networks in Saccharomyces cerevisiae. Science 298:799–804
36. Xie Z, Hu S, Qian J, Blackshaw S, Zhu H (2011) Systematic characterization of protein-DNA interactions. Cell Mol Life Sci 68: 1657–1668
37. Farnham PJ (2009) Insights from genomic profiling of transcription factors. Nat Rev Genet 10:605–616
38. Cawley S, Bekiranov S, Ng HH, Kapranov P, Sekinger EA, Kampa D et al (2004) Unbiased mapping of transcription factor binding sites along human chromosomes 21 and 22 points to widespread regulation of noncoding RNAs. Cell 116:499–509
39. Kapranov P, Cawley SE, Drenkow J, Bekiranov S, Strausberg RL, Fodor SP, Gingeras TR (2002) Large-scale transcriptional activity in chromosomes 21 and 22. Science 296: 916–919
40. Kampa D, Cheng J, Kapranov P, Yamanaka M, Brubaker S, Cawley S et al (2004) Novel RNAs identified from an in-depth analysis of the transcriptome of human chromosomes 21 and 22. Genome Res 14:331–342
41. Kapranov P, Cheng J, Dike S, Nix DA, Duttagupta R, Willingham AT et al (2007) RNA maps reveal new RNA classes and a possible function for pervasive transcription. Science 316:1484–1488
42. Spencer WC, Zeller G, Watson JD, Henz SR, Watkins KL, McWhirter RD et al (2011) A spatial and temporal map of C. elegans gene expression. Genome Res 21:325–341
43. Stolc V, Gauhar Z, Mason C, Halasz G, van Batenburg MF, Rifkin SA et al (2004) A gene expression map for the euchromatic genome of Drosophila melanogaster. Science 306:655–660
44. Eichner J, Zeller G, Laubinger S, Rätsch G (2011) Support vector machines-based identification of alternative splicing in Arabidopsis thaliana from whole-genome tiling arrays. BMC Bioinformatics 12:55
45. Sebat J, Lakshmi B, Troge J, Alexander J, Young J, Lundin P et al (2004) Large-scale copy number polymorphism in the human genome. Science 305:525–528
46. Brommesson S, Jönsson G, Strand C, Grabau D, Malmström P, Ringnér M, Fernö M, Hedenfalk I (2008) Tiling array-CGH for the assessment of genomic similarities among synchronous unilateral and bilateral invasive breast cancer tumor pairs. BMC Clin Pathol 8:6
47. Shinawi M, Cheung SW (2008) The array CGH and its clinical applications. Drug Discov Today 13:760–770
48. Zhang X, Yazaki J, Sundaresan A, Cokus S, Chan SW, Chen H et al (2006) Genome-wide high-resolution mapping and functional analysis of DNA methylation in arabidopsis. Cell 126:1189–1201
49. Zilberman D, Gehring M, Tran RK, Ballinger T, Henikoff S (2007) Genome-wide analysis of Arabidopsis thaliana DNA methylation uncovers an interdependence between methylation and transcription. Nat Genet 39:61–69
50. Rauch TA, Wu X, Zhong X, Riggs AD, Pfeifer GP (2009) A human B cell methylome at 100-base pair resolution. Proc Natl Acad Sci USA 106:671–678
51. Bernstein BE, Kamal M, Lindblad-Toh K, Bekiranov S, Bailey DK, Huebert DJ et al (2005) Genomic maps and comparative analysis of histone modifications in human and mouse. Cell 120:169–181
52. Otto C, Reiche K, Hackermuller J (2012) Detection of differentially expressed segments in tiling array data. Bioinformatics 28:1471–1479
53. Johnson WE, Li W, Meyer CA, Gottardo R, Carroll JS, Brown M, Liu XS (2006) Model-based analysis of tiling-arrays for ChIP-chip. Proc Natl Acad Sci USA 103:12457–12462
54. Kadener S, Rodriguez J, Abruzzi KC, Khodor YL, Sugino K, Marr MT 2nd, Nelson S, Rosbash M (2009) Genome-wide identification of targets of the drosha-pasha/DGCR8 complex. RNA 15:537–545
55. Lee SJ, Trostel A, Le P, Harinarayanan R, Fitzgerald PC, Adhya S (2009) Cellular stress

created by intermediary metabolite imbalances. Proc Natl Acad Sci USA 106: 19515–19520

56. Judy JT, Ji H (2009) TileProbe: modeling tiling array probe effects using publicly available data. Bioinformatics 25:2369–2375
57. Bieda M, Xu X, Singer MA, Green R, Farnham PJ (2006) Unbiased location analysis of E2F1-binding sites suggests a widespread role for E2F1 in the human genome. Genome Res 16:595–605
58. Zhang ZD, Rozowsky J, Lam HY, Du J, Snyder M, Gerstein M (2007) Tilescope: online analysis pipeline for high-density tiling microarray data. Genome Biol 8:R81
59. Zacher B, Kuan PF, Tresch A (2010) Starr: simple tiling ARRay analysis of Affymetrix ChIP-chip data. BMC Bioinformatics 11:194
60. Bailey TL, Williams N, Misleh C, Li WW (2006) MEME: discovering and analyzing DNA and protein sequence motifs. Nucleic Acids Res 34:W369–W373
61. Okoniewski MJ, Yates T, Dibben S, Miller CJ (2007) An annotation infrastructure for the analysis and interpretation of Affymetrix exon array data. Genome Biol 8:R79
62. Sorek R, Cossart P (2010) Prokaryotic transcriptomics: a new view on regulation, physiology and pathogenicity. Nat Rev Genet 11:9–16
63. Sasidharan R, Agarwal A, Rozowsky J, Gerstein M (2009) An approach to comparing tiling array and high throughput sequencing technologies for genomic transcript mapping. BMC Res Notes 2:150
64. Agarwal A, Koppstein D, Rozowsky J, Sboner A, Habegger L, Hillier LW et al (2010) Comparison and calibration of transcriptome data from RNA-Seq and tiling arrays. BMC Genomics 11:383
65. Bradford JR, Hey Y, Yates T, Li Y, Pepper SD, Miller CJ (2010) A comparison of massively parallel nucleotide sequencing with oligonucleotide microarrays for global transcription profiling. BMC Genomics 11:282

Part II

Transcriptional Regulation and Expression

Chapter 2

Design of Tiling Arrays and Their Application to Bacterial Transcriptome Analysis

Takeshi Akama, Kazuaki Nakamura, Akito Tanoue, and Koichi Suzuki

Abstract

Whole-genome sequencing and annotation have clarified total gene number and structure in a variety of organisms. Microarrays have facilitated transcriptome analysis through the use of probes that target a large number of genes based on genomic information. However, microarrays are limited in that they can only examine known or predicted genes; non-annotated genes and noncoding regions cannot be accounted for.

Recent advances in technology have led to the design of tiling arrays, which contain a vastly increased number of spotted probes, and at higher density. Tiling arrays cover the entire genome of a prokaryotic species in an unbiased fashion by designing a large number of probes. Upon hybridization of total RNA, all the transcribed regions of the genome, irrespective of gene annotation, can be detected. As opposed to next-generation sequencing, tiling arrays are cost-effective, easy to analyze, and have been used for experiments as diverse as transcriptome analysis, ChIP-chip, and DNA sequence variation detection. In this chapter, the methods for bacterial tiling array slide design, RNA sample preparation, hybridization, and data analysis are described.

Key words Tiling array, Transcriptome, Bacteria, Genome, Noncoding region

1 Introduction

Microarray techniques permit the analysis of hundreds of thousands of genes in a single experiment by using probes mounted on a glass slide. Microarrays spotted with probes specific for every gene have been developed and applied as a high-throughput technique for analyzing gene expression in a variety of organisms [1–5]. However, only known genes are taken into account with probe design; other regions of the genome, including non-annotated coding regions, pseudogenes, and noncoding regions, cannot be analyzed in an open reading frame (ORF) array (Fig. 1). Noncoding regions, which do not encode functional proteins, were previously thought to be "junk" DNA and were not analyzed. However, there has been increased awareness of the functions of noncoding

Tin-Lap Lee and Alfred Chun Shui Luk (eds.), *Tiling Arrays: Methods and Protocols*, Methods in Molecular Biology, vol. 1067,
DOI 10.1007/978-1-62703-607-8_2, © Springer Science+Business Media New York 2013

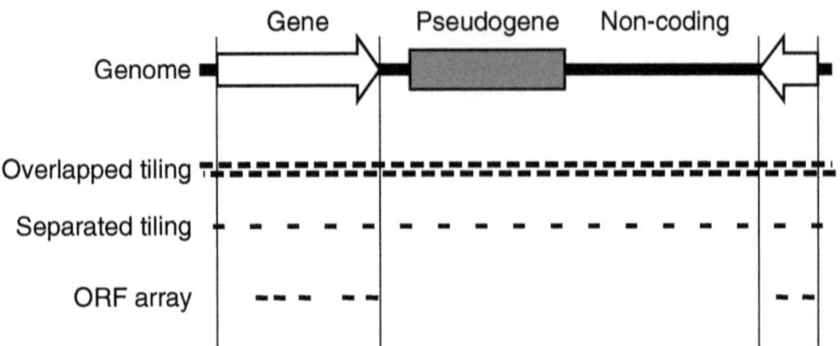

Fig. 1 Probe arrangement of the tiling array and ORF array. For the genome sequence of the target organism, probes for the tiling array are arranged at equal intervals, although the ORF array only probes known genes. Thus, not only the coding regions but also pseudogenes and other noncoding regions from the whole genome are covered by the tiled probes in a tiling array. Depending on the genome size and maximum probe number of the array slide, neighboring probes can be overlapping or separate

regions, and some pseudogenes are known to be expressed and function in the regulation of gene expression [6–8].

In tiling arrays, probes are designed to span the entire genome of an organism without bias, regardless of whether or not the regions are annotated (Fig. 1). With typical eukaryotic genomes, the tiling array is designed for each genome contig because of the large genome size. Tiling arrays for prokaryotes, which have smaller sized genomes compared to eukaryotes, can be designed to cover the whole genome in an unbiased arrangement and thus can analyze the entire transcriptome. It has been applied to a variety of species such as *Bacillus subtilis* [9], *Neisseria meningitidis* [10], and *Saccharomyces cerevisiae* [11]. Previously, a whole-genome tiling array was designed for *Mycobacterium leprae* (*M. leprae*), a bacterium with a small genome and a large number of noncoding regions, and its transcriptome was analyzed [12]. This application of the tiling array for *M. leprae* is depicted as an example.

2 Materials

2.1 Tiling Array Design

1. Genomic DNA sequence: National Center for Biotechnology Information (http://www.ncbi.nlm.nih.gov/genome/browse/) or Genomes OnLine Database (GOLD) (http://www.genomesonline.org/cgi-bin/GOLD/index.cgi).
2. 1× 4.2 M high density HD4 array (Roche NimbleGen, Madison, WI).
3. Custom array synthesis system (Roche NimbleGen).

2.2 RNA Extraction from Bacteria

1. RNA Protect Bacteria Reagent (Qiagen, Germantown, MD).
2. 1.0 mm Zirconia Beads (BioSpec Products, Bartlesville, OK).

3. *mir*Vana miRNA Isolation Kit (Ambion, Austin, TX).
4. Micro Smash Homogenizer (TOMY, Tokyo Japan).
5. DNase I (TaKaRa, Kyoto Japan).

2.3 Preparation of Labeled Double-Stranded DNA

1. SuperScript II (Life Technologies, Carlsbad, CA).
2. RNaseOUT (Life Technologies).
3. RNase A (Novagen, Madison, WI).
4. Cy3-9mer Wobble primer (TriLink Biotechnologies, San Diego, CA).
5. Klenow fragment (New England Biolabs, Ipswich, MA).

2.4 Hybridization and Analysis of the Tiling Array

1. NimbleGen Hybridization Kit (NimbleGen Systems, Madison, WI).
2. MAUI Hybridization System (BioMicro Systems, Salt Lake City, UT).
3. NimbleGen Wash Buffer Kit (NimbleGen Systems).
4. Microarray High-Speed Centrifuge (Arrayit, Sunnyvale, CA).
5. GenePix 4000B scanner (Molecular Devices, Sunnyvale, CA).
6. NIMBLESCAN 2.3 (NimbleGen Systems).

2.5 Quantitative Real-Time PCR

1. Primer-BLAST (http://www.ncbi.nlm.nih.gov/tools/primer-blast/), Primer3 (http://www.bioinformatics.nl/cgi-bin/primer3plus/primer3plus.cgi), or other primer designing programs.
2. FastStart Universal SYBR Green Master (Rox) (Roche Diagnostics, Indianapolis, IN).
3. MicroAmp Optical 96-well Reaction Plate (Life Technologies).
4. MicroAmp Optical Adhesive Film (Life Technologies).
5. ABI Prism 7000 Sequence Detection System (Life Technologies).

3 Methods

The tiling array can identify unexpected transcriptional activity in a number of genomic regions, although the data obtained may have a lower signal-to-noise ratio. This is due to the fact that the probe sequences cannot be designed based on DNA specificity or GC content. Thus, proper probe length is important to improve the signal-to-noise ratio. The interval (degree of overlap) of the probes, which is decided by the genome size of the target organism and the number of probes mounted on the array slide, determines the resolution of expressed regions on the genome (Fig. 1). RNA degradation must be low during sample preparation as short segments of degraded RNA can hybridize nonspecifically.

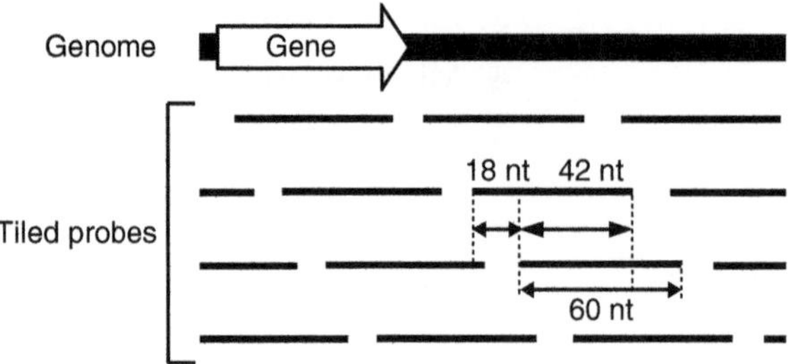

Fig. 2 Interval and overlap of the tiling array probe. For each of the two adjacent probes, 18 nucleotides of the 60-mer probe are shifted, resulting in an overlap of 42 nucleotides. Four consecutive probes share a partial sequence

3.1 Tiling Array Design

1. The genome sequence of the organism to be analyzed can be obtained from the NCBI Web site, "Genomic DNA sequence." The Web site GOLD can also be available to search both complete and incomplete genome projects. In case annotation is lacking, the genome sequence can be analyzed using the annotation programs on the GOLD Web site.
2. The genome sequence is divided into equal segments for probe design. The length of each probe is 60-mer and adjacent probes are spaced 18 nucleotides apart, resulting in probes that overlap every 42 nucleotides on the genome (Fig. 2) (*see* **Note 1**). Because the maximum allowable probe number is 4,200,000 in the Roche Nimblegen custom array, probe length and overlap between adjacent probes can be determined according to the specific genome length (*see* **Note 2**).
3. Random probes and control probes are also designed. Multiple random probes that have different sequences should be prepared to determine the background level. Control probes are hybridized with control nucleotides (Alignment Oligo of NimbleGen Hybridization Kit), spiked into samples, and used as an internal control to compare data obtained from different arrays.
4. Based on the above designs, the probes are synthesized on an array slide by Roche Nimblegen (*see* **Note 3**).

3.2 RNA Extraction from Bacteria

1. Prepare bacterial cells yielding at least 1 μg RNA upon extraction. Suspend the cells in 2 ml of RNA Protect Bacteria Reagent and incubate for 10 min at room temperature. Centrifuge the cell suspension at $9{,}000 \times g$ for 6 min at 4 °C and remove the supernatant.
2. Resuspend the cell pellet in a mixture of 2 ml of RNA Protect Bacteria Reagent, 0.4 ml of 1.0 mm zirconia beads (*see* **Note 4**), and 0.6 ml of lysis/binding buffer from the *mir*Vana miRNA Isolation Kit (*see* **Note 5**).

3. Homogenize the mixture four times at 3,000 rpm for 3 min using a Micro Smash Homogenizer followed by a freeze–thaw cycle. Repeat four times.
4. Add 60 μl (0.1 volume of lysis/binding buffer) of homogenate additive to the homogenate and vortex. After a 10-min incubation on ice, add 600 μl (one volume) of acid–phenol:chloroform and vortex for 1 min. Centrifuge the sample at maximum speed for 5 min at room temperature to separate the organic and aqueous phases.
5. Preheat the elution solution to 95 °C at this time.
6. After centrifugation, transfer the supernatant to a new tube, add 750 μl (1.25 volumes) of ethanol, and mix at room temperature. Transfer the mixture (up to 700 μl at a time) to a filter cartridge that has been placed in a collection tube.
7. Centrifuge the cartridge at 10,000 × *g* for 15 s and discard the flow through. Repeat the centrifugation step in order to pass all the mixture through the cartridge.
8. Apply 700 μl of wash solution 1 to the filter cartridge and centrifuge at 10,000 × *g* for 10 s. Discard the flow through.
9. Apply 500 μl of wash solution 2/3 and centrifuge at 10,000 × *g* for 10 s. Discard the flow through. Repeat this step.
10. Centrifuge the cartridge at 10,000 × *g* for 10 s for 1 min and transfer the filter cartridge to a new collection tube.
11. Apply 100 μl of preheated (95 °C) elution solution to the center of the filter. Centrifuge the filter cartridge at maximum speed for 30 s.
12. Measure the absorbance of the eluent at 260 and 280 nm to determine RNA concentration and purity. The ratio of A_{260}/A_{280} should be greater than 1.8.
13. Dispense one or more μg of RNA into a microtube and add 30 μl (0.1 volume) of DNase I buffer, 15 U of DNase I, and distilled water to a total volume of 300 μl. Incubate the mixture at 37 °C for 30 min followed by acid–phenol:chloroform extraction and ethanol precipitation. Dissolve the RNA pellet in 10 μl RNase-free water (*see* **Note 6**).

3.3 Preparation of Labeled Double-Stranded DNA

1. Mix up to 5 μg of total RNA with 250 ng random primer, 1 μl of 10 mM dNTP mix, and RNase-free water to a total volume of 12 μl. Incubate the mixture at 65 °C for 5 min and immediately transfer to ice.
2. Add 4 μl of 5× first-strand buffer, 2 μl of 0.1 M DTT, and 1 μl of RNaseOUT to the mixture. Incubate at 25 °C for 2 min.

3. Add 1 μl of SuperScript II RT to the mixture, and incubate at 25 °C for 10 min. Next, incubate the mixture at 42 °C for 50 min followed by heating at 70 °C for 15 min.
4. Incubate the synthesized cDNA from **step 3** with 10 ng of RNase A at 37 °C for 10 min followed by phenol:chloroform extraction and ethanol precipitation.
5. Incubate 1 μg of ds-cDNA at 98 °C for 10 min with 1 OD_{600} unit of Cy3-9mer Wobble primer (*see* **Note** 7) and distilled water to a total volume of 80 μl.
6. Add 10 μl of dNTPs (10 mM each) and 2 μl (100 U) of Klenow fragment and incubate at 37 °C for 2 h. Stop the reaction by adding 10 μl (0.1 volume) of 0.5 M EDTA, and precipitate the labeled cDNA with isopropanol.

3.4 Hybridization of the Tiling Array

1. Warm the MAUI Hybridization System to 42 °C for 3 h prior to use.
2. Resuspend the Cy3-labeled sample in 1.7 μl of distilled water plus 3.3 μl of sample tracking control, 9.0 μl of 2× NimbleGen Hybridization Buffer, 3.6 μl Hybridization Component A, and 0.37 μl of Alignment Oligo.
3. Denature the mixture by heating at 95 °C for 5 min and maintain at 42 °C until sample loading.
4. Remove an X1 Mixer from its package 30 min prior to use. Using the Precision Mixer Alignment Tool (PMAT), place the tiling array slide on the mixer.
5. Place the mixer-slide assembly in the slide bay of the MAUI Hybridization System. Apply the sample through the fill port until it starts to leak from the vent port. Remove excess sample that has leaked from the mixer and adhere a mixer port seal over the fill and the vent ports.
6. Hybridize the samples for 18 h at 42 °C using mix mode B of the MAUI Hybridization System.
7. Prepare wash I, II, and III by mixing 225 ml of distilled water, 25 μl of 1 M DTT, and 10× wash buffer I, II, or III.
8. Load the mixer-slide assembly into the Mixer Disassembly Tool immersed in wash I at 42 °C. After peeling off the mixer, agitate the slide for 10–15 s.
9. Transfer the slide to a processing container containing wash I. Shake the processing container at least 20 times every 10 s. After 2 min, transfer the slide into wash II and agitate for 1 min. Next, transfer the slide into wash III and agitate for 15 s.
10. Remove the slide from the wash III and dry for 1 min using the microarray high-speed centrifugation. Prior to scanning, keep the slide shielded from light.

3.5 Scanning of the Array Slide

1. Turn on the GenePix 4000B scanner and launch the GenePix software 10 min prior to use.
2. Blow compressed gas across the slide to remove any contaminants. Place the slide in the slide carriage so that the array is face down and the barcode end is closest to you.
3. Move the black lever on the left side of the carriage to the left until the slide is lying flat in the carriage. Release the lever so that the slide is gently pushed to the right side of the carriage and held firmly in place.
4. Close the slide carriage and shut the scanner door.
5. On the Hardware Settings dialog box of GenePix software, select the following settings: Wavelength to 532, PMT gain to 500, Power to 100, Pixel size to 5, Lines to average to one, and Focus position to zero.
6. Click the Scan Area icon and draw a box that surrounds the array image by dragging the mouse. The box should be slightly larger than the array image.
7. Click the play icon to scan the array slide.
8. Adjust the PMT gain so that normalized counts at the 65,000-intensity level are 1e-5. Stop the scan and restart under the new setting.
9. Save the obtained image. Remove the array slide from the scanner and store in a dark desiccator for rescanning.
10. Launch the NimbleScan software. The scanned image, a design file (.ndf), and a gene description file (.ngd) will open.
11. Perform the auto brightness/contrast adjust function and the auto align tool.
12. Adjust the grid so that the corner fiducial controls line up correctly with the grid.
13. Run the local alignment tool to fine-tune the alignment and save the file.
14. Create a pair file by selecting Reports-Pair in the Analysis menu, which contains the raw data and lists the probe intensities of the array. The gridded image file is assigned the same . ndf file as the one in **step 10**.
15. Create a general feature format file (.gff) by selecting Reports-GFF in the Analysis menu, which can be opened with the SignalMap software. A position file (.pac) will be assigned (*see* **Note 8**).

3.6 Data Analysis on SignalMap Software

1. Launch the SignalMap software and the tiling array data file will open with a gene annotation file.
2. The positive regions of hybridized signals can be detected by applying the Find Peaks function in the Track menu. Shorter

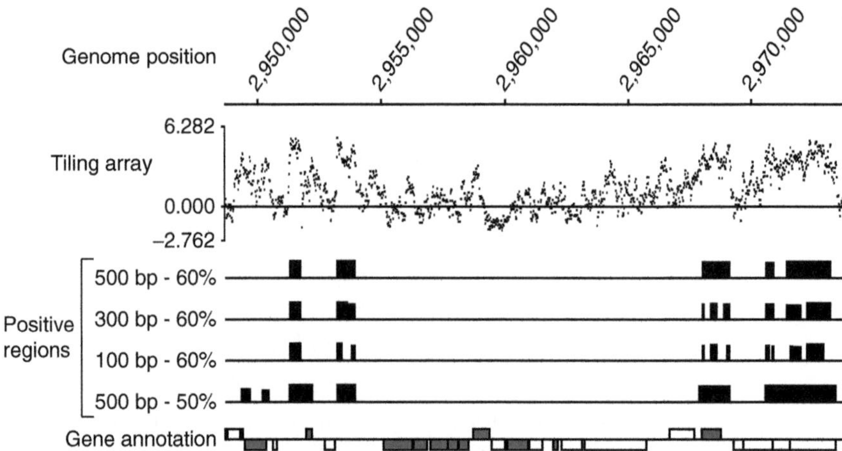

Fig. 3 Detection of positive regions from various thresholds. The data from the 60-mer tiling array are analyzed for positive regions by using the Find Peaks menu in the SignalMap software. Peak window size is changed from 500 to 100 bp, and the Percent of Peak Threshold is changed to 60 or 50 %. By limiting the Peak window size, shorter positive regions can be detected, although some positive regions may be fragmented. A lower Percent of Peak Threshold increases the sensitivity, as shown on the genome position around 2,950,000. The positive regions on a noncoding segment of the genome (before 2,955,000) and on a pseudogene (before 2,970,000) are shown

positive regions can be detected by decreasing the Peak window size, although some positive regions may be fragmented. Percent of Peak Threshold affects the sensitivity. Positive regions that have lower signal intensity can be detected by decreasing the Percent of Peak Threshold (Fig. 3).

3. In the line created after applying the Find Peaks function, positive regions are shown as rectangles. The number of Rows indicates the number of positive regions. The precise range and highest score of the positive region are shown when hovering the mouse over the rectangles.

3.7 Quantitative Real-Time PCR

1. The cDNA used for the tiling array can also be subjected to real-time PCR analysis.
2. The primers can be designed using Primer-BLAST, Primer3, or other primer design programs. Primers are designed to amplify 80–300 bp PCR products due to the amplification efficiency of target genes.
3. To generate mean and standard error values of the relative cDNA amounts, prepare the reactions in triplicate. For one of the reactions, mix the cDNA with 10 μl of 2× FastStart Universal SYBR Green Master (Rox), 0.04 μl of 100 μM forward and reverse primer, and distilled water to a total volume of 20 μl. Furthermore, prepare the reactions in duplicate using a series of diluted sample for the standard cDNA to draw standard curve.

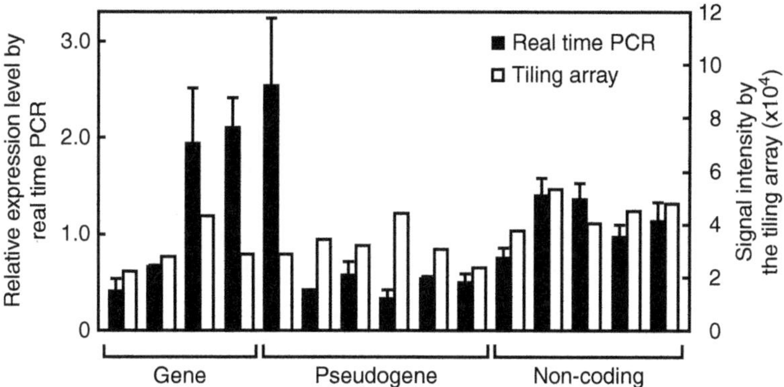

Fig. 4 Comparison of RNA expression between real-time PCR and the tiling array. For positive regions in genes, pseudogenes, and other noncoding regions, relative RNA expression levels detected by quantitative real-time PCR (*left vertical axis*) are compared with the signal intensities detected by the tiling array (*right vertical axis*). All of the positive regions in the tiling array can be detected by real-time PCR; however, the magnitude of their expression levels differs with the result from the tiling array

4. Apply the mixtures to a MicroAmp Optical 96-well Reaction Plate and seal the top of the plate with MicroAmp Optical Adhesive Film.
5. Turn on ABI PRISM 7000, the computer, and launch the ABI PRISM 7000 SDS software.
6. Create a new document for Absolute Quantification and set detectors which are selected SYBR in the reporter dye menu for each of the PCR products in the Detector Manager window.
7. Define the Sample name, Detector, Task, and quantity options in the Well Inspector dialog box on the Plate tab. The same sample names are filled for triplicate samples. For the standard curve, mixtures of a series of diluted sample are selected in the Task option to STD and dilution factors are filled in the Quantity option.
8. Under the Instrument tab, set the sample volume to 20 μl and check the box labeled Dissociation Protocol. Use the default protocol setting on the Thermal Cycler: an initial step of 2 min at 50 °C and 10 min at 95 °C followed by 40 cycles of 15 s at 95 °C and 1 min at 60 °C.
9. After saving the file, place an Optical Cover Compression Pad on the top of the plate and set the plate to ABI PRISM 7000. Close the lid of the machine and start the run.
10. When the run is finished, check the Standard Curve and Dissociation curve. To certify that there are no PCR by-products, the Dissociation Curve should have single peak at the same position among triplicates. After exporting the results, the relative amounts of cDNA may be obtained. Examples of the comparison between the quantitative real-time PCR result and the tiling array result are shown in Fig. 4.

4 Notes

1. Longer probes have higher specificity, but lower resolution. Although 29-mer probes are spotted for some predesigned ORF arrays, they give vaguer signal patterns than 60-mer probes. By aligning the tiling array data and ORF array data, it is shown that specific signals can be obtained by the tiling array with 60-mer probes like ORF array as shown in Fig. 5.
2. Probes can be designed to both the plus and minus strand, as long as the total probe number is under half of the maximum of the custom array slide. In this design, probe specificity can be checked with the data from both strands.
3. Tiling Array Tools (http://www.affymetrix.com/partners_programs/programs/developer/TilingArrayTools/index.affx) (Affymetrix, Santa Clara, CA) can also be used.
4. For some bacterial species, lysozyme treatment may be used instead of homogenization. Resuspend the cell pellet in 2 ml of RNA Protect Bacteria Reagent, add 0.4 ml of 1.2 mg/ml lysozyme in TE, and incubate for 5 min at room temperature. Centrifuge the mixture at 5,000×*g* for 10 min. Repeat the lysozyme incubation step once more, and add 0.6 ml lysis/binding buffer from the *mir*Vana miRNA Isolation Kit. The remaining steps are the same from Subheading 3.2, **step 4**.
5. The mirVana miRNA Isolation Kit is used to extract total RNA including short RNA. Other RNA extraction reagents like TRIZOL (Life Technologies, Carlsbad, CA) can also be used.
6. If an excess amount of RNA has been extracted, checking for RNA degradation by electrophoresis is recommended. Prepare samples for loading by mixing several μg of total RNA with 2.5 μl of 10× MOPS (MERCK, Darmstadt, Germany), 1 μl of 1.0 mg/ml ethidium bromide (Sigma), 4.5 μl formaldehyde, and 12.5 μl formamide. Incubate the mixture at 65 °C for 15 min and chill on ice. Add 2.5 μl of RNA gel loading solution (Quality Biological, Inc., Gaithersburg, MD), and load the samples on a 1 % agarose gel with 1× MOPS running buffer. An optimal ratio of 23S/16S rRNA amount should be greater than 1.0.
7. Cy3-labeled random nonamer may be used instead of Cy3-9mer Wobble primer. Dissolve 1 O.D. of primer in 42 μl of 12.5 mM Tris–HCl, pH7.4, 1.25 mM $MgCl_2$, and 25 mM β-mercaptoethanol.
8. The values of signal intensity are supplied as raw data in the .gff file. To convert them to log2 values, open the .gff file in Microsoft Excel and use the LOG function. The modified file can be saved as a tab-delimited text file and can be changed from .txt to .gff.

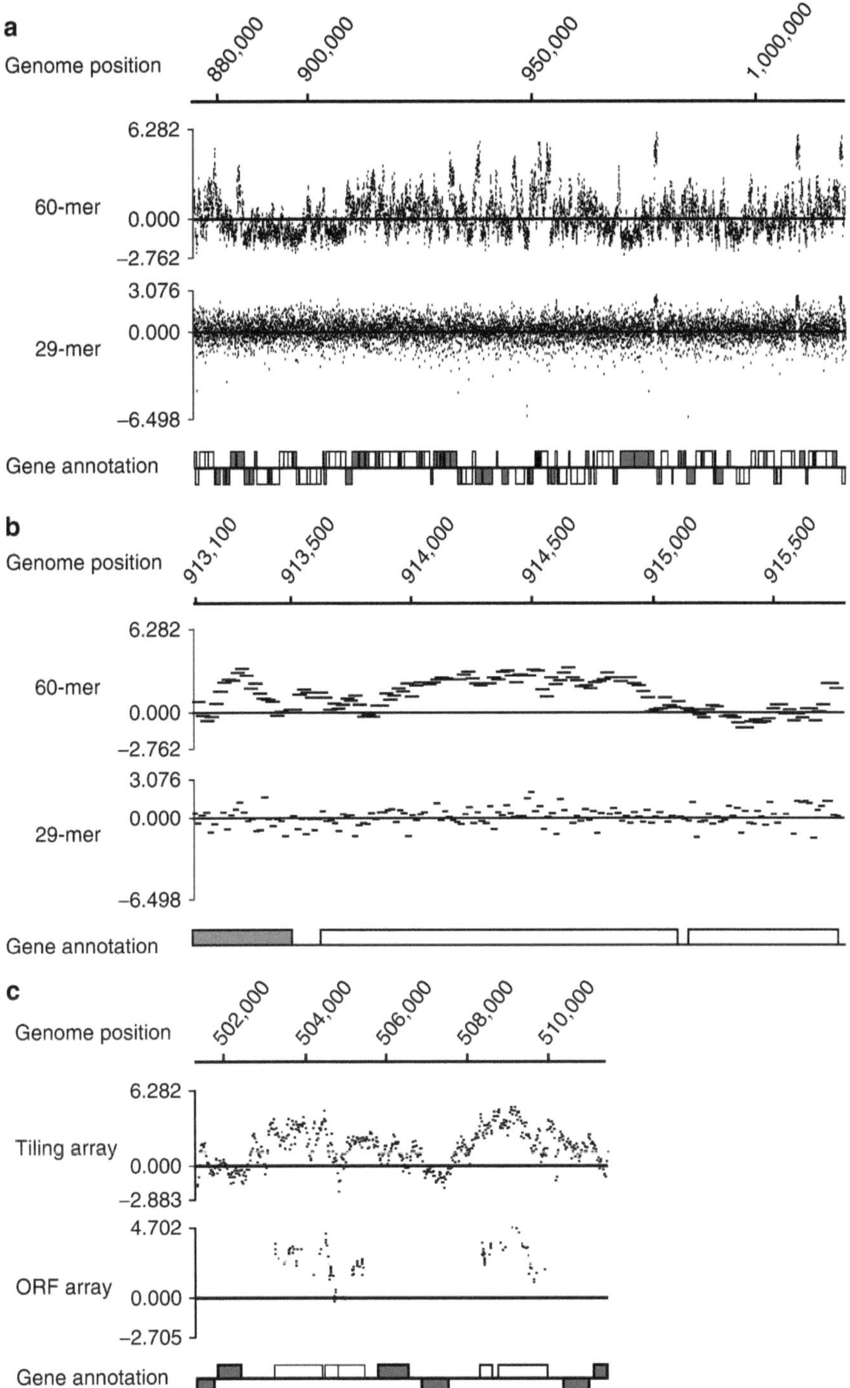

Fig. 5 Influence of probe length on signal specificity. For *M. leprae* genomic DNA (GENBANK Acc. No. NC_002677), the tiling arrays spotted with 60- or 29-mer probes were designed. The obtained signal intensity is graphed on the vertical axis and each probe is horizontally arrayed against the corresponding genome sequence using the SignalMap software. The data from a 60-mer array are shown in the *upper position*, and the data from a 29-mer array are shown in the *middle position. White rectangles* positioned *underneath* indicate the range of coding regions and *gray rectangles* indicate the range of pseudogenes. Broad (**a**) and narrow (**b**) ranges of the genome are shown. (**c**) The data from the 60-mer tiling array are aligned with the data from the ORF array. In the ORF array, different 20 probes are designed against one coding region to ensure sequence specificity

References

1. Franchini AG, Egli T (2006) Global gene expression in *Escherichia coli* K-12 during short-term and long-term adaptation to glucose-limited continuous culture conditions. Microbiology 152:2111–2127
2. Niehus E, Gressmann H, Ye F, Schlapbach R, Dehio M, Dehio C et al (2004) Genome-wide analysis of transcriptional hierarchy and feedback regulation in the flagellar system of *Helicobacter pylori*. Mol Microbiol 52: 947–961
3. Zhou D, Han Y, Qiu J, Qin L, Guo Z, Wang X et al (2006) Genome-wide transcriptional response of *Yersinia pestis* to stressful conditions simulating phagolysosomal environments. Microbes Infect 8:2669–2678
4. Stewart GR, Wernisch L, Stabler R, Mangan JA, Hinds J, Laing KG et al (2002) Dissection of the heat-shock response in *Mycobacterium tuberculosis* using mutants and microarrays. Microbiology 148:3129–3138
5. Akama T, Tanigawa K, Kawashima A, Wu H, Ishii N, Suzuki K (2010) Analysis of *Mycobacterium leprae* gene expression using DNA microarray. Microb Pathog 49:181–185
6. Hirotsune S, Yoshida N, Chen A, Garrett L, Sugiyama F, Takahashi S et al (2003) An expressed pseudogene regulates the messenger-RNA stability of its homologous coding gene. Nature 423:91–96
7. Lin H, Shabbir A, Molnar M, Lee T (2007) Stem cell regulatory function mediated by expression of a novel mouse Oct4 pseudogene. Biochem Biophys Res Commun 355:111–116
8. Suzuki K, Nakata N, Bang PD, Ishii N, Makino M (2006) High-level expression of pseudogenes in *Mycobacterium leprae*. FEMS Microbiol Lett 259:208–214
9. Rukmana A, Morimoto T, Takahashi H, Giyanto, Ogasawara N (2009) Assessment of transcriptional responses of Bacillus subtilis cells to the antibiotic enduracidin, which interferes with cell wall synthesis, using a high-density tiling chip. Genes Genet Syst 84:253–267
10. Swiderek H, Claus H, Frosch M, Vogel U (2005) Evaluation of custom-made DNA microarrays for multilocus sequence typing of Neisseria meningitidis. Int J Med Microbiol 295:39–45
11. Perocchi F, Xu Z, Clauder-Munster S, Steinmetz LM (2007) Antisense artifacts in transcriptome microarray experiments are resolved by actinomycin D. Nucleic Acids Res 35:e128
12. Akama T, Suzuki K, Tanigawa K, Kawashima A, Wu H, Nakata N et al (2009) Whole-genome tiling array analysis of Mycobacterium leprae RNA reveals high expression of pseudogenes and noncoding regions. J Bacteriol 191:3321–3327

Chapter 3

Transcript Profiling in Arabidopsis with Genome Tiling Microarrays

Diana Coman, Wilhelm Gruissem, and Lars Hennig

Abstract

Microarray technology is at present a standardized workflow for genome-wide expression analysis. Whole-genome tiling microarrays have emerged as an important platform for flexible and comprehensive expression profiling. In this chapter we describe a detailed standardized workflow for experiments assessing the transcriptome of Arabidopsis using tiling arrays and provide useful hints for critical steps from experimental design to data analysis. Although the protocol is optimized for AGRONOMICS1 arrays, it can readily be adapted to other tiling arrays. AGRONOMICS1 is the first platform that enables strand-specific expression analysis of the Arabidopsis genome with a single array. Moreover, it includes all perfect match probes from the original ATH1 array, allowing readily integration with the large existing ATH1 knowledge base. This workflow is designed for the analysis of raw data for any number of samples and it does not pose any particular hardware requirements.

Key words AGRONOMICS1, Arabidopsis, Genome tiling array, Microarray, Transcriptome

1 Introduction

Over the past 15 years microarray technology rapidly evolved and has matured to routine and standardized workflows for genome-wide expression analysis in many organisms including all major model and crop plants [1–3]. One general purpose of gene expression microarrays is to capture a snapshot of expression levels of many or all genes under a particular condition by probing the transcripts present in the cell, tissue, or organ under investigation. The detection principle of microarrays is based on the complementarity of the four bases in the nucleic acids. Short sequences such as artificially synthesized oligonucleotide probes or PCR products are fixed on the array surface and bind complementary regions of the labeled RNA sample. Direct detection, if RNA was fluorescently labeled, or indirect detection, if labeled RNA is probed with a fluorescently labeled antibody, generates a fluorescence signal whose

Tin-Lap Lee and Alfred Chun Shui Luk (eds.), *Tiling Arrays: Methods and Protocols*, Methods in Molecular Biology, vol. 1067, DOI 10.1007/978-1-62703-607-8_3, © Springer Science+Business Media New York 2013

intensity is proportional to the amount of mRNA present in the sample. After recording signal intensities, the downstream workflow involves various bioinformatics approaches, e.g., identification of differentially expressed genes or common expression patterns, detection of alternative splicing, or identification of novel transcripts.

For the model plant *Arabidopsis thaliana*, Affymetrix ATH1 microarrays are most widely used [4]. ATH1 arrays measure gene expression based on probe sets consisting of 11 perfect match and corresponding 11 mismatch oligonucleotide probes that are complementary to target mRNA sequences. However, only genes that were annotated at the time of array design in the year 2004 are interrogated implying that expression of about one-third of the currently annotated genes present in the Arabidopsis reference strain is missed.

In contrast, high-density whole-genome or genome tiling arrays have probes that are tiled to cover the whole genome or spaced at regular intervals, and allow unbiased detection of transcripts from the entire genome [5]. Whole-genome tiling microarrays (WGAs) have emerged as an important platform with usability going beyond simple gene expression profiling [1]. This type of array can be used for diverse applications ranging from genome-wide gene model-independent expression analysis [6–9], discovery of novel transcribed loci [10–12], and alternative splicing analysis [13] to identification of transcription factor binding sites using ChIP-chip [14, 15], high-resolution mapping of DNA methylation sites, and histone modifications [16–18].

For whole-genome transcriptome analysis in Arabidopsis, several commercially available tiling array platforms are available such as the Affymetrix GeneChip Arabidopsis tiling array set, the Affymetrix AGRONOMICS1 tiling array, and the NimbleGen *A. thaliana* ChIP-chip 385K Whole-Genome Tiling Set. Figure 1 compares the design features of the three mentioned WGAs for Arabidopsis.

The Affymetrix GeneChip Arabidopsis tiling array set comprises a pair of single arrays representing the forward strand (1.0F) and the reverse strand (1.0R), respectively (http://www.affymetrix.com). Each array contains over 3.2 million perfect match and mismatch probe pairs (~6.4 million probes in total) tiled with 35 bp spacing throughout the complete non-repetitive Arabidopsis genome. This tiling array is designed for high-specificity target recognition owed to the consistent probe length of 25 nucleotides. This type of WGA has been used to profile the transcriptome of various tissues and developmental stages of Arabidopsis plants and/or under different conditions [6–9].

The NimbleGen *A. thaliana* ChIP-chip 385K Whole-Genome Tiling Set contains three arrays that together cover all unique regions of the Arabidopsis genome with probes spaced every 90 bp.

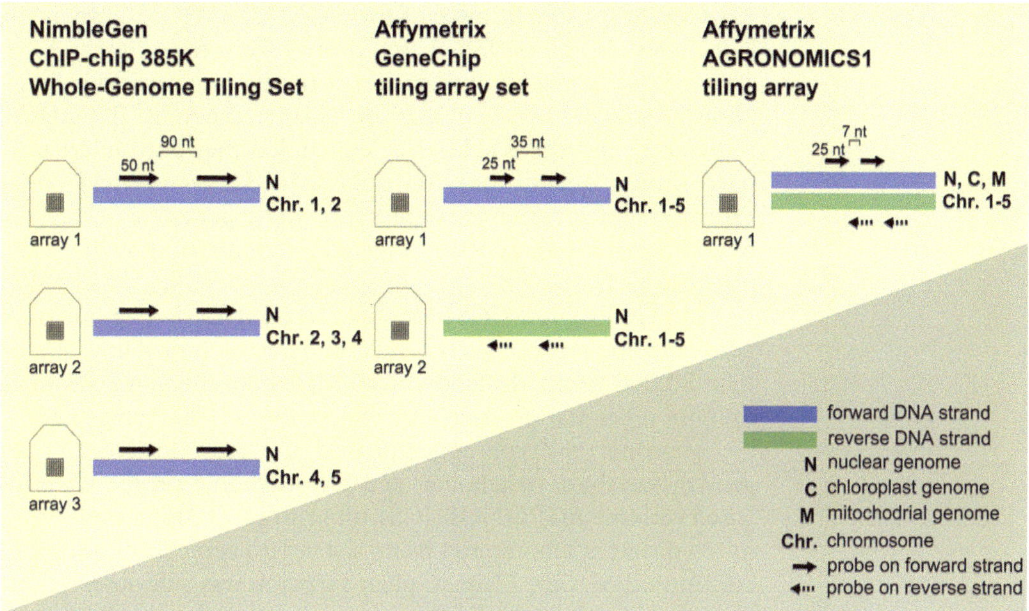

Fig. 1 Schematic Tiling WGA design of the NimbleGen *A. thaliana* ChIP-chip 385K Whole-Genome Tiling Set, Affymetrix GeneChip Arabidopsis tiling array set, and Affymetrix AGRONOMICS1 tiling array

This tiling array type comprises isothermal probes, i.e., the length of the probe is variable in favor of consistent melting temperature, which enables higher overall signal levels due to increased sequence available for target hybridization (http://www.nimblegen.com). However, this design sacrifices feature density and coverage resulting in overall lower resolution in comparison to the Affymetrix tiling platforms. This type of WGA was successfully used for studies of epigenetic modifications in Arabidopsis [19–21]. The single-strand array design makes strand-specific gene expression profiling impossible on a single chip.

The AGRONOMICS1 tiling array was developed by the European AGRON-OMICS research consortium (http://www.agron-omics.eu) in collaboration with Affymetrix. The design of this new tiling array includes probes representing both strands of the entire Arabidopsis genome with one 25-mer probe per 35 bp genome sequence window on average [22]. AGRONOMICS1 is the first WGA that enables strand-specific expression analysis of the Arabidopsis genome with a single array. The design of the AGRONOMICS1 array is described in detail elsewhere [22]. More than 90 % of the annotated nuclear genes are represented by at least three specific probes on the AGRONOMICS1 array. In addition, the mitochondrial and chloroplast genomes are completely represented. Probes with more than one perfect match and probes positioned in repetitive sequences are excluded [22].

One of the key features of the AGRONOMICS1 array is that it includes also all perfect match probes from the original ATH1 array, allowing readily integration with the large existing ATH1 knowledge base [23]. A comparison between the AGRONOMICS1 tiling array and the ATH1 array concluded that both microarrays have similar quality, but AGRONOMICS1 arrays yield robust expression information for significantly more genes, i.e., expression of additional 7,646 genes is detected in comparison to the ATH1 platform [22].

The AGRONOMICS1 array can be used for diverse functional genomics applications such as reliable expression profiling of more than 30,000 genes, detection of alternative splicing, and chromatin immunoprecipitation coupled to microarrays (ChIP-chip) [22].

Massonet and colleagues probed the reproducibility of leaf growth variables, metabolite, and transcriptome profiles (interrogated with the AGRONOMICS1 tiling array) of Arabidopsis grown in ten different laboratories using a standardized cultivation protocol. Although some plant-to-plant variation was still observed, the use of a strictly standardized experimental workflow in all laboratories enabled consistent and comparable results [24]. By using AGRONOMICS1 tiling arrays to assess the transcriptome profile of Toc159-deficient plants, which are disturbed in plastidial protein import, and by subsequent integration with proteomics data, the consequences of systemic transcriptional responses were delimited from impaired protein import resulting from loss of photosynthetic capacity [25]. Andriankaja and colleagues used the AGRONOMICS1 tiling array to characterize the progression of the third leaf of Arabidopsis through the transition from primary to secondary morphogenesis. Of the total number of differentially expressed genes identified in this study with the AGRONOMICS1 array, 18.9 % were not present on the Affymetrix ATH1 arrays [26].

Microarray analyses provide insights into the transcriptome within single experiments (e.g., wild-type versus mutant or control versus treatment) but also open up broad perspectives to further metadata studies by integrating expression data from various sources collected in curated repositories (e.g., Genevestigator®) (Table 1). In this respect, the AGRONOMICS1 tiling array design is of great value for the Arabidopsis community allowing integration with other microarray studies, in particular those that used ATH1 microarrays and together profiled over 7,000 samples [23].

New transcriptomics technologies are continuously developed, and examples include nCounter, OpenArray, Fluidigm, and RNA-seq as contemporary alternatives to microarrays [27–30]. Among these, RNA-seq poses a particular serious challenge to microarrays. Both microarrays and sequencing have their own specific biases, which in turn affect the ability to accurately measure differential gene expression. For example, cross-hybridization and sequence content of the probes are well known to affect expression

Table 1
Downstream microarray data analysis

Method	Algorithm	Principle	Software, Web site
Differential expression	Limma (Linear Models for Microarray Data)	Fits a linear model to the expression data for each probe set	R, Bioconductor; http://www.bioconductor.org
Antisense transcripts	Custom CDF	Analysis of signals from probes targeting the antisense strand	R; www.agron-omics.eu
Alternative splicing	FIRMA	Unsupervised detection of differential splicing based on the RMA model	R, AROMA; http://www.aroma-project.org/
Dimension reduction	Principal Component Analysis (PCA)	Targets directions in the data space along maximum variance to visualize and identify general data groupings and outliers	R, MeV; http://www.tm4.org/mev/
Clustering	k means, Partition Around Medioids (PAM), hierarchical clustering	More accurate groupings and better characterization of gene expression patterns	R, MeV; http://www.tm4.org/mev/
Co-expression	ATTEDII, Genevestigator	Identifies coregulated transcripts across multiple experiments	http://atted.jp; https://www.genevestigator.com/
Ontology	GOslim, Gene Set Enrichment Analysis (GSEA)	Assigns genes to pathways and attempts to make associations with biological processes, molecular functions, and cellular components, using controlled vocabularies	http://www.geneontology.org/GO.tools.microarrayshtml; http://www.broadinstitute.org/gsea/index.jsp

measures on microarray platforms [31, 32]. Concurrently, a dependency between transcript length and differential expression was observed in the RNA-seq data along with a GC bias and mapping ambiguity for paralogous sequences [33, 34]. Nevertheless, because microarrays and sequencing offer complementary advantages they will likely coexist in the near future [35].

The flexibility, robustness, and reproducibility of the microarray technology and data analysis workflows in conjunction with high throughput at lower costs will most probably support their

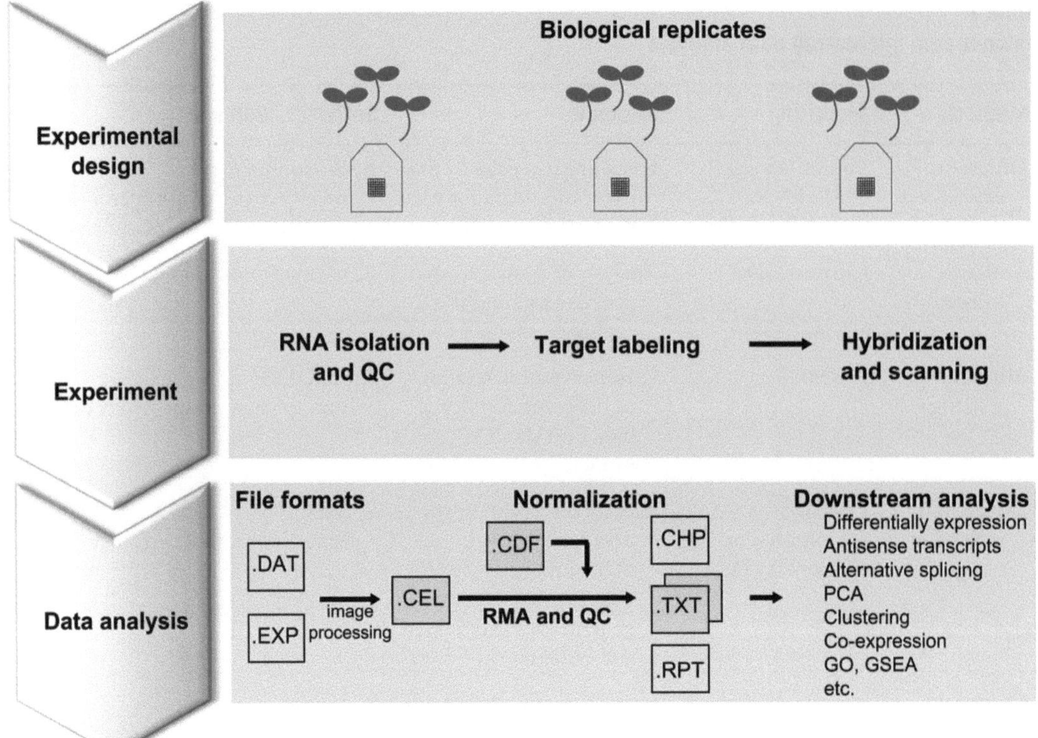

Fig. 2 Standard workflow for transcriptomics microarray experiments using the AGRONOMICS1 array

extensive use in the future for routine transcriptome experiments. To minimize error and variability between laboratories and across microarray platforms, standardized workflows (i.e., plant growth, sample collection, RNA isolation and labeling, hybridization, and data analysis) should be followed by the plant community as much as possible [24, 36, 37].

In this chapter we describe a detailed standardized workflow (Fig. 2) for experiments assessing the transcriptome of Arabidopsis using the AGRONOMICS1 tiling array and provide useful hints for critical steps from experimental design to data analysis.

2 Materials

1. Plant material (biological replicates *see* **Note 1**).
2. Plant RNeasy MiniKit (Qiagen) for RNA isolation (*see* **Note 2**).
3. NanoDrop ND1000 (NanoDrop Technologies) and Bioanalyzer 2100 (Agilent) for RNA integrity and quality control (*see* **Note 3**).
4. AGRONOMICS1 microarray (Affymetrix) (*see* **Note 4**).

5. GeneChip 3' IVT Express Kit (Affymetrix) for microarray target labeling (*see* **Note 5**).
6. Hybridization station and scanner for image capture (Affymetrix) (*see* **Note 6**).
7. Software for data analysis: Freely available R, AROMA, and AGRONOMICS1 scripts (*see* **Note 7**).

3 Methods

3.1 RNA Isolation and Quality Control

1. Isolate total RNA from the plant material with the Plant RNeasy MiniKit, according to the manufacturer's instructions (*see* **Note 2**).
2. The quality of the RNA is essential, thus, it is mandatory to assess the purity and integrity of the isolated total RNA (*see* **Note 3**). Use only samples with 260 nm/280 nm ratios between 1.8 and 2.1, measured for instance with a NanoDrop ND1000, and 28S/18S ratios between 1.5 and 2 measured with a Bioanalyzer 2100.

3.2 Microarray Target Preparation (cDNA from mRNA and Labeling)

1. Use the 3' IVT Express Kit following the manufacturer's instructions to prepare the microarray targets (*see* **Note 5**). In brief, use the 3' IVT Express Kit with 50 ng of total RNA (*see* **Note 8**). First, perform the reverse transcription of the total RNA into the first-strand cDNA, which will be further converted into a double-stranded DNA template for transcription. Subsequently, label the double-stranded cDNA by in vitro transcription in the presence of biotin-labeled ribonucleotides generating biotinylated cRNA (*see* **Note 9**). It is advisable to include the poly(A) controls as recommended by the manufacturer.
2. Test the quality and quantity of the biotinylated cRNA by measuring the absorbance at 260 nm and 280 nm (*see* **Note 10**).
3. In addition to the cRNA yield, the size distribution can be accurately evaluated using a Bioanalyzer with the RNA 6000 Nano Kit.
4. Fragment the biotinylated cRNA samples (15 μg) in fragmentation buffer at 94 °C to obtain random sizes of cRNA ranging from 35 bp to 200 bp (*see* **Note 11**).

3.3 Array Hybridization and Data Acquisition

1. Mix the fragmented biotinylated cRNA samples and the hybridization mix with the hybridization control and Control Oligo B2 to make a final volume of 300 μl according to the manufacturer's recommendations.
2. Hybridize the samples onto AGRONOMICS1 arrays at 45 °C for 16 h to facilitate sequence-specific binding.

3. Wash the arrays in a Fluidics Station 450 using the FS450_0004 protocol.
4. Scan the arrays with the Affymetrix GeneChip Scanner 3000 (*see* **Notes 6** and **12**).

3.4 Data Analysis

1. To analyze the data, follow the workflow below using freely available R packages and scripts (*see* **Notes 7** and **13**). This workflow performs the normalization and generates the probe set summaries and is designed for the analysis of raw data generated from an AGRONOMICS1 tiling array for an arbitrary number of samples. It does not pose any particular hardware requirements. The procedure can also be used for other Affymetrix tiling arrays if appropriate Chip Definition Format (CDF) files are available.
2. Collect the following files required for the analysis:
   (a) The fluorescence intensities files (.CEL files): sample_1.CEL, sample_2.CEL, and sample_3.CEL (*see* **Note 12**).
   (b) The CDF file: agronomics1_TAIR10_gene.cdf (*see* **Note 14**).
   (c) The actual analysis script: agronomicsTools10.r script (*see* **Notes 7** and **15**).
3. Generate the specific folder structure required for the AROMA package. Store the data files (.CEL), the annotation (.CDF) file, and the agronomicsTools10.r script for data analysis within specifically named folders (*see* Fig. 3 for the required folder structure and **Note 16**).
4. Run the script below within the R environment to transform the probe intensities into gene expression values. This script uses the AGRONOMICS1_RMA function defined in the agronomicsTools10.r file. The text in green represents comments to the respective R commands:

```
# set the working directory
    > setwd("X:\\aroma")
# load the source code (contains the function
  for normalization, flagging algorithm,
  probe set summary calculation)
    > source("Scripts\\agronomicsTools10.r")
# call the standard RMA analysis using the
  agronomics1_TAIR10_gene.cdf annotation file
    > AGRONOMICS1_RMA(experimentName)
```

   In brief, the AROMA package implemented in R takes as input .CEL files and the specified CDF file and matches the probe intensity from the .CEL files with the genome

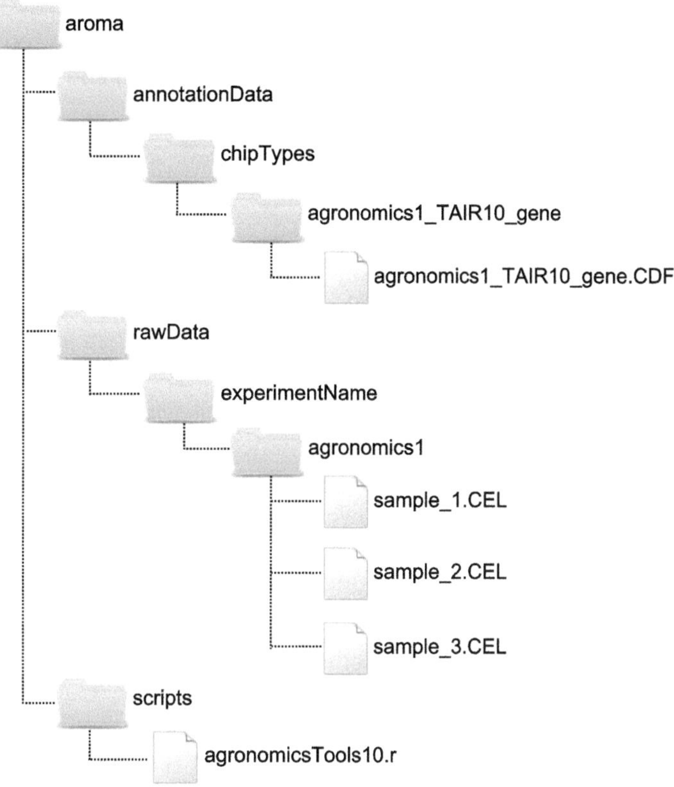

Fig. 3 Standard AROMA folder structure used for data analysis with the agronomics1Tools10.r script

annotation data from the CDF file, for which each probe had been mapped to the Arabidopsis genome. Poorly performing probes (signal intensity <2^5 in all samples) are flagged and excluded from further analysis as described in Rehrauer et al. 2010 (*see* **Note 17**). The flexibility and adaptability of this workflow working with custom CDFs and a flagging algorithm may be useful, for example, when the expression is assessed for Arabidopsis accessions other than the Col0 reference strain, which was used to design the array (*see* **Note 18**). Next, based on the popular robust multi-array average (RMA) method [38] the background correction, normalization, and calculation of probe set summaries are performed (*see* **Note 19**). At the end of this analysis, a new folder is automatically added to the aroma folder structure. This folder contains the result files that will be used for further downstream analysis (*see* Table 1).

5. Register your experiment in a public data repository (e.g., Gene Expression Omnibus—GEO at NCBI or ArrayExpress at EBI) in compliance with the MIAME format (*see* **Note 20**).

6. Perform downstream data analysis. Several orienting examples for further data analysis are given in Table 1 along with their principle, potential use, and availability.

4 Notes

1. When designing an experiment, at least three independent biological replicates are strongly advised to enable estimation of biological variance. Technical replicates are not necessary unless the study investigates technical variability of microarrays or labeling protocols. To buffer plant-to-plant variation, it is advisable to pool material from many individual plants.
2. The Plant RNeasy MiniKit (Qiagen) was successfully used to isolate total RNA from diverse plant material such as seedlings, flowers, and leaves [22, 24–26]. Alternative protocols and kits can be used, for instance the TRIzol® reagent (Invitrogen). Nonetheless, more specialized plant material may need optimized methods [39, 40].
3. High-quality total RNA is needed as starting material. RNA samples should not be contaminated with cellular material, proteins, DNA, and compounds associated with the RNA isolation procedure such as ethanol, phenol, or salts. DNA and protein contaminations can interfere with accurate measurement of RNA concentration, whereas organic solvent contamination can affect the efficiency of the reverse transcription and subsequently reduce the amplification levels. We recommend to use the NanoDrop ND1000 for assessing the RNA quality and the Bioanalyzer 2100 to determine the RNA integrity. To accurately evaluate the RNA integrity (the proportion of RNA that is full length), the RNA integrity number (RIN) can be calculated as described by Agilent (www.chem.agilent.com).
4. The AGRONOMICS1 tiling array design is registered at ArrayExpress (A-AFFY-155) and is commercially available through Affymetrix.
5. The 3' IVT Express Kit was compared with two alternative labeling methods (GeneChip One-Cycle cDNA synthesis kit, Affymetrix, and WT Ovation Pico System, NuGEN) [22]. The 3' IVT Express Kit required 50 ng RNA, whereas the GeneChip One-Cycle needed 1 μg RNA and both had increased strand specificity. The WT Ovation Pico System did not reliably distinguish between strands. Recently, an alternative random labeling kit, the GeneChip Whole Transcript (WT) Sense Target Labeling Assay (Affymetrix), was shown to outperform the classical 3' IVT Express Kit by reducing 3' bias and increasing sensitivity for long and organelle transcripts [41].

6. Different hybridization stations and scanners can be used but must be compatible with the microarray platform. The AGRONOMICS1 tiling array was successfully used with the Affymetrix Fluidics Station 450 and the Affymetrix GeneChip Scanner 3000, which are also routinely used for hybridization of ATH1 and genome-wide tiling arrays.
7. Library files and scripts for data analysis of the AGRONOMICS1 tiling array are freely available at http://www.agron-omics.eu/index.php/resource_center/tiling-array and at www.slu.se/genetics/resources/agronomics1.
8. The minimum amount of the total RNA for the 3' IVT Express Kit is 50 ng, and the recommended amount is 100 ng of total RNA. The optimal total RNA input and the incubation time of the IVT reaction are sample dependent. Thus, within a given experiment the same amount of input RNA should be used and the IVT incubation time should be kept constant.
9. The 3' IVT Express Kit protocol includes an amplification step in the form of an in vitro transcription to synthesize biotin-labeled cRNA; multiple copies of biotin-modified cRNA are created from the double-stranded cDNA templates.
10. The cRNA concentration may vary depending on the amount and quality of poly(A) RNA in the input total RNA. The cRNA can be concentrated by centrifugation under vacuum at medium–low temperature. Complete drying of the cRNA samples should be avoided. Purified cRNA can be stored up to 1 year at −20 °C (avoid repeated freeze–thaw cycles to minimize to cRNA degradation).
11. To ensure optimal assay sensitivity, it is recommended to work with small volumes during the cRNA fragmentation step to minimize the magnesium concentration in the final hybridization mix. The fragmented cRNA should be used immediately after synthesis or store undiluted, at −20 °C (or −70 °C for longer term storage).
12. The recorded fluorescence intensities for each probe are stored in CEL files, and thus a given sample with three replicates has the specific CEL file set: sample_1.CEL, sample_2.CEL, and sample_3.CEL.
13. Software for data analysis: The R environment for statistical computing is required [42]. Alternatively, the analysis can be performed using the Affymetrix Power Tools (APT). However, it is recommended to use the AROMA package [43] implemented in R to make full use of the flagging algorithm.
14. The custom CDF files were obtained by mapping a .bpmap file (specifies the layout of a microarray, i.e., the type, annotation, the spatial order of oligonucleotides, and the probe sequences on the array) to a reference genome (the mapping depends

on the used genome release version of The Arabidopsis Information Resource—TAIR). Particularly, in the agronomics1_TAIR10_gene.cdf file every probe set matches exactly one gene according to the latest Arabidopsis annotation available at TAIR10. Other types of custom CDF files are available at http://www.agron-omics.eu/index.php/resource_center/tiling-array (e.g., exon-specific and previous TAIR releases for backward compatibility).

15. The agronomicsTools10.r script defines two functions: AGRONOMICS1_RMA and aromaRMA. AGRONOMICS1_RMA is a wrapper to perform RMA analysis specifically with AGRONOMICS1 arrays. This function includes the probe filtering algorithm described by Rehrauer and colleagues [22]. aromaRMA is a generic wrapper for RMA analysis of any Affymetrix array with any matching CDF file. Both functions require the aroma.affymetrix package and its special directory structure to organize raw, annotation and processed data.
16. The AROMA folder structure can be generated with a helper script found at http://www.agron-omics.eu/index.php/resource_center/tiling-array. The folder structure must be precisely followed. The file and folder names are case sensitive. For example, "Sample_1" is not the same as "sample_1." More details can be found at http://www.aroma-project.org/.
17. The flagging algorithm identifies and removes probes with signal intensity lower than 5 on log2 scale. Technically, the flagged probes are only removed in memory from the loaded CDF. Next, for any probe sets that contain less than three probes, the flagged probes with the highest mean signals are repeatedly unflagged until at least three unflagged probes are kept in the probe set [22].
18. The flagging algorithm can be used to identify and/or exclude probes that do not hybridize optimally due to sequence polymorphism between different Arabidopsis accessions.
19. The normalization step is required to remove noise intrinsic to technical errors and to obtain comparable data. The central step of the RMA analysis is the quantile normalization. Each quantile value, including minimum, median, and maximum, is replaced by its average across the respective quantiles of all arrays, rendering identical data distributions across arrays [38]. A quality control analysis, using for example the aroma.affymetrix package, is recommended to identify possible outlier samples (i.e., arrays) indicating technical failure or unusual variability in the biological replicates.
20. MIAME stands for *M*inimum *I*nformation *A*bout a *M*icroarray *E*xperiment and describes the standard information for recording and reporting microarray-based gene expression data [44, 45].

Acknowledgments

We thank our colleagues at the Functional Genomics Center Zurich for helpful discussions. Research in the authors' laboratories is supported by the Sixth and Seventh Framework Programs of the European Commission through the AGRON-OMICS integrated project (grant no. LSHG–CT–2006–037704) to W.G., L.H., and others and the TiMet collaborative project (grant 245143) to W.G., and by a grant from the Swedish Research Council to L.H.

References

1. Mockler TC, Chan S, Sundaresan A, Chen H, Jacobsen SE, Ecker JR (2005) Applications of DNA tiling arrays for whole-genome analysis. Genomics 85:1–15
2. Galbraith DW (2006) DNA microarray analyses in higher plants. OMICS 10:455–473
3. Galbraith DW, Edwards J (2010) Applications of microarrays for crop improvement: here, there, and everywhere. Bioscience 60:337–348
4. Redman JC, Haas BJ, Tanimoto G, Town CD (2004) Development and evaluation of an Arabidopsis whole genome Affymetrix probe array. Plant J 38:545–561
5. Bertone P, Trifonov V, Rozowsky JS, Schubert F, Emanuelsson O, Karro J et al (2006) Design optimization methods for genomic DNA tiling arrays. Genome Res 16:271–281
6. Yamada K, Lim J, Dale JM, Chen H, Shinn P, Palm CJ et al (2003) Empirical analysis of transcriptional activity in the Arabidopsis genome. Science 302:842–846
7. Matsui A, Ishida J, Morosawa T, Mochizuki Y, Kaminuma E, Endo TA et al (2008) Arabidopsis transcriptome analysis under drought, cold, high-salinity and ABA treatment conditions using a tiling array. Plant Cell Physiol 49:1135–1149
8. Hazen SP, Naef F, Quisel T, Gendron JM, Chen H, Ecker JR et al (2009) Exploring the transcriptional landscape of plant circadian rhythms using genome tiling arrays. Genome Biol 10:R17
9. Laubinger S, Zeller G, Henz SR, Sachsenberg T, Widmer CK, Naouar N et al (2008) At-TAX: a whole genome tiling array resource for developmental expression analysis and transcript identification in *Arabidopsis thaliana*. Genome Biol 9:R112
10. Stolc V, Samanta MP, Tongprasit W, Sethi H, Liang S, Nelson DC et al (2005) Identification of transcribed sequences in *Arabidopsis thaliana* by using high-resolution genome tiling arrays. Proc Natl Acad Sci U S A 102: 4453–4458
11. Hanada K, Zhang X, Borevitz JO, Li WH, Shiu SH (2007) A large number of novel coding small open reading frames in the intergenic regions of the *Arabidopsis thaliana* genome are transcribed and/or under purifying selection. Genome Res 17:632–640
12. Swiezewski S, Crevillen P, Liu F, Ecker JR, Jerzmanowski A, Dean C (2007) Small RNA-mediated chromatin silencing directed to the 3' region of the Arabidopsis gene encoding the regulator, FLC. Proc Natl Acad Sci U S A 104:3633–3638
13. Ner-Gaon H, Fluhr R (2006) Whole-genome microarray in Arabidopsis facilitates global analysis of retained introns. DNA Res 13: 111–121
14. Thibaud-Nissen F, Wu H, Richmond T, Redman JC, Johnson C, Green R (2006) Development of Arabidopsis whole-genome microarrays and their application to the discovery of binding sites for the TGA2 transcription factor in salicylic acid-treated plants. Plant J 47:152–162
15. Turck F, Roudier F, Farrona S, Martin-Magniette ML, Guillaume E, Buisine N et al (2007) Arabidopsis TFL2/LHP1 specifically associates with genes marked by trimethylation of histone H3 lysine 27. PLoS Genet 3:e86
16. Martienssen RA, Doerge RW, Colot V (2005) Epigenomic mapping in Arabidopsis using tiling microarrays. Chromosome Res 13: 299–308
17. Zhang X, Yazaki J, Sundaresan A, Cokus S, Chan SW, Chen H et al (2006) Genome-wide high-resolution mapping and functional analysis of DNA methylation in Arabidopsis. Cell 126:1189–1201

18. Li X, Wang X, He K, Ma Y, Su N, He H et al (2008) High-resolution mapping of epigenetic modifications of the rice genome uncovers interplay between DNA methylation, histone methylation, and gene expression. Plant Cell 20:259–276
19. Inagaki S, Miura-Kamio A, Nakamura Y, Lu F, Cui X, Cao X et al (2010) Autocatalytic differentiation of epigenetic modifications within the Arabidopsis genome. EMBO J 29:3496–3506
20. Farrona S, Thorpe FL, Engelhorn J, Adrian J, Dong X, Sarid-Krebs L et al (2011) Tissue-specific expression of FLOWERING LOCUS T in Arabidopsis is maintained independently of Polycomb Group protein repression. Plant Cell 23:3204–3214
21. Moghaddam AMB, Roudier F, Seifert M, Bérard C, Magniette MLM, Ashtiyani RK et al (2011) Additive inheritance of histone modifications in *Arabidopsis thaliana* intra-specific hybrids. Plant J 67:691–700
22. Rehrauer H, Aquino C, Gruissem W, Henz SR, Hilson P, Laubinger S et al (2010) AGRONOMICS1: a new resource for Arabidopsis transcriptome profiling. Plant Physiol 152:487–499
23. Hruz T, Laule O, Szabo G, Wessendorp F, Bleuler S, Oertle L et al (2008) Genevestigator V3: a reference expression database for the meta-analysis of transcriptomes. Adv Bioinformatics 2008:1–5
24. Massonnet C, Vile D, Fabre J, Hannah MA, Caldana C, Lisec J et al (2010) Probing the reproducibility of leaf growth and molecular phenotypes: a comparison of three Arabidopsis accessions cultivated in ten laboratories. Plant Physiol 152:2142–2157
25. Bischof S, Baerenfaller K, Wildhaber T, Troesch R, Vidi PA, Roschitzki B et al (2011) Plastid proteome assembly without Toc159: photosynthetic protein import and accumulation of N-acetylated plastid precursor proteins. Plant Cell 23:3911–3928
26. Andriankaja M, Dhondt S, De Bodt S, Vanhaeren H, Coppens F, De Milde L et al (2012) Exit from proliferation during leaf development in *Arabidopsis thaliana*: a not-so-gradual process. Dev Cell 22:64–78
27. Geiss GK, Bumgarner R, Birditt B, Dahl T, Dowidar N, Dunaway DL et al (2008) Direct multiplexed measurement of gene expression with color-coded probe pairs. Nat Biotechnol 26:317–325
28. Spurgeon SL, Jones RC, Ramakrishnan R (2008) High throughput gene expression measurement with real time PCR in a microfluidic dynamic array. PLoS One 3:e1662
29. Weaver S, Dube S, Mir A, Qin J, Sun G, Ramakrishnan R et al (2010) Taking qPCR to a higher level: analysis of CNV reveals the power of high throughput qPCR to enhance quantitative resolution. Methods 50:271–276
30. Wang Z, Gerstein M, Snyder M (2009) RNA-Seq: a revolutionary tool for transcriptomics. Nat Rev Genet 10:57–63
31. Naef F, Magnasco MO (2003) Solving the riddle of the bright mismatches: labeling and effective binding in oligonucleotide arrays. Phys Rev E Stat Nonlin Soft Matter Phys 68:011906
32. Wu Z, Irizarry RA (2010) Stochastic models inspired by hybridization theory for short oligonucleotide arrays. J Comput Biol 12:882–893
33. Bullard JH, Purdom E, Hansen KD, Dudoit S (2010) Evaluation of statistical methods for normalization and differential expression in mRNA-Seq experiments. BMC Bioinformatics 11:94
34. Oshlack A, Wakefield MJ (2009) Transcript length bias in RNA-seq data confounds systems biology. Biol Direct 4:14
35. Wheelan SJ, Murillo FM, Boeke JD (2008) The incredible shrinking world of DNA microarrays. Mol Biosyst 4:726–732
36. Irizarry RA, Warren D, Spencer F, Kim IF, Biswal S, Frank BC et al (2005) Multiple-laboratory comparison of microarray platforms. Nat Methods 2:345–350
37. Larkin JE, Frank BC, Gavras H, Sultana R, Quackenbush J (2005) Independence and reproducibility across microarray platforms. Nat Methods 2:337–344
38. Irizarry RA, Hobbs B, Collin F, Beazer-Barclay YD, Antonellis KJ, Scherf U, Speed TP (2003) Exploration, normalization, and summaries of high density oligonucleotide array probe level data. Biostatistics 4:249–264
39. Wuest SE, Vijverberg K, Schmidt A, Weiss M, Gheyselinck J, Lohr M et al (2010) Arabidopsis female gametophyte gene expression map reveals similarities between plant and animal gametes. Curr Biol 20:506–512
40. Schmidt A, Schmid MW, Grossniklaus U (2012) Analysis of plant germline development by high-throughput RNA profiling: technical advances and new insights. Plant J 70:18–29
41. Mueller M, Patrignani A, Rehrauer H, Gruissem W, Hennig L (2012) Evaluation of alternative RNA labeling protocols for

transcript profiling with Arabidopsis AGRONOMICS1 tiling arrays. Plant Methods 8:18

42. R Development Core Team (2012) R: a language and environment for statistical Computing. R Foundation for Statistical Computing, Vienna, Austria. ISBN 3-900051-07-0, URL http://www.R-project.org/
43. Bengtsson H, Simpson K, Bullard J, Hansen K (2008) Aroma.affymetrix: a generic framework in R for analyzing small to very large Affymetrix data sets in bounded memory, Tech Report #745, Department of Statistics, University of California, Berkeley, CA.
44. Brazma A, Hingamp P, Quackenbush J, Sherlock G, Spellman P, Stoeckert C et al (2001) Minimum information about a microarray experiment (MIAME)-toward standards for microarray data. Nat Genet 29:365–371
45. Zimmermann P, Schildknecht B, Craigon D, Garcia-Hernandez M, Gruissem W, May S et al (2006) MIAME/Plant – adding value to plant microarrray experiments. Plant Methods 2:1

Chapter 4

Genome-Wide Analysis of Transcription Factor-Binding Sites in Skeletal Muscle Cells Using ChIP-Seq

Chung-Il An and Nobuko Hagiwara

Abstract

Transcriptional regulation of gene expression constitutes a fundamental mechanism of many developmental processes. Therefore, identification and characterization of binding sites of transcription factors are important for uncovering the mechanisms of a particular developmental process. Here, we describe detailed procedures for genome-wide analysis of binding sites of a transcription factor involved in the fiber-type differentiation of skeletal muscle. By conducting ChIP-seq followed by a series of downstream analyses, in-depth information on binding sites of transcription factors can be obtained in a genome-wide manner.

Key words ChIP-seq, Sox6, Skeletal muscle differentiation, Primary myotubes, Library preparation, Alignment, Peak calling, Motif discovery

1 Introduction

Every developmental process is tightly controlled, in part, by a number of transcription factors coordinating the regulation of genes involved in development. Therefore, discovering and characterizing the transcription factor-binding sites can provide mechanistic insight into a particular developmental process through identification of the target genes regulated by the factors.

DNA–protein interactions in vivo can be detected by a method called chromatin immunoprecipitation (ChIP) [1]. With this approach, DNA fragments bound by transcription factors or other DNA-binding proteins can be enriched by using specific antibodies to proteins of interest. With the introduction of microarray techniques, it became possible to expand this approach to a much larger scale: hybridization of DNA fragments obtained by ChIP to high-density tiling arrays allowed identification of DNA–protein interactions at a genome-wide scale (ChIP-chip) [2]. Recent advances in high-throughput sequencing technology now present

Tin-Lap Lee and Alfred Chun Shui Luk (eds.), *Tiling Arrays: Methods and Protocols*, Methods in Molecular Biology, vol. 1067,
DOI 10.1007/978-1-62703-607-8_4, © Springer Science+Business Media New York 2013

another option for looking into genome-wide transcription factor-binding sites (ChIP-seq) [3]. Therefore, two options are now available for genome-wide mapping of DNA–protein interactions in vivo: ChIP-chip (ChIP followed by tiling array) and ChiP-Seq (ChIP followed by high-throughput sequencing). The first, ChIP-chip, is particularly well suited for investigations involving a specific region of the genome (e.g., a specific chromosome or a particular large stretch of DNA). However, due to the intrinsic nature of microarray technologies, some substantial limitations are present in this ChIP-chip approach such as low-resolution, cross-hybridization between similar sequences, and narrow dynamic range [4, 5].

The recently developed ChIP-seq technique overcame these problems and therefore can provide a more detailed and reliable genome-wide analysis [4, 5]. At this stage, neither technique is perfect in that ChIP-seq can miss a small number of binding sites which can be detected with ChIP-chip [6]. However, especially at the genome-wide level, ChIP-seq has considerable advantages compared to its predecessor technique, and a recent improvement enabled even single-nucleotide resolution [7]. In our laboratory, we wanted to understand the mechanisms involved in skeletal muscle fiber-type differentiation and how the transcription factor Sox6 impacts fiber-type gene regulation on a genome-wide level. Therefore, we applied ChIP-seq for genome-wide mapping of Sox6-binding sites on the mouse genome in skeletal muscle cells [8].

Sox6 is a member of the evolutionarily conserved Sox transcription factor family and is involved in many developmental processes [9–12]. In developing skeletal muscle, the loss of a functional Sox6 protein leads to upregulation of multiple slow fiber-specific genes [13, 14], indicating that Sox6 functions as a transcriptional suppressor of slow fiber-specific genes during muscle development. To understand the molecular mechanisms of fiber-type differentiation of skeletal muscle regulated by Sox6, identifying Sox6 target genes was necessary. To achieve this goal, we took advantage of the ChIP-seq technique to unravel genome-wide Sox6-binding sites in skeletal muscle cells [8].

Here, we describe detailed procedures of the Sox6 ChIP-seq analysis using mouse fetal skeletal myotubes. The procedures include six major steps: sample preparation, ChIP, quantitative PCR (qPCR), library preparation, sequencing, and computational data analysis which consists of three major sub-steps including sequence alignment, peak calling, and downstream analysis (e.g., peak annotation and motif discovery) (Fig. 1). Among these steps, sample preparation, ChIP, qPCR, and downstream analysis are common to both ChIP-chip and ChIP-seq experiments (Fig. 1). Therefore, the procedures for these steps described in this chapter can be readily applicable to ChIP-chip study. The major differences

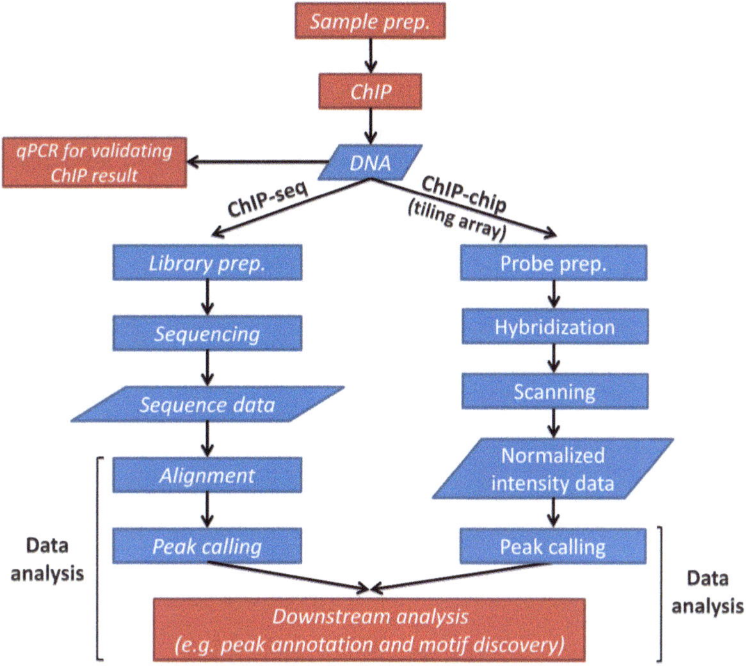

Fig. 1 Workflow of the two genome-wide methods for analyzing protein–DNA interactions in vivo. Sample preparation, ChIP, qPCR, and downstream analyses steps are common to both the methods. After the ChIP step, ChIP-seq uses high-throughput sequencing while ChIP-chip uses microarrays to obtain sequence information on ChIP-enriched DNA. Common steps between ChIP-seq and ChIP-chip are shown in *red*, and the ChIP-seq workflow covered in this chapter is shown in *italics*

between ChIP-seq and ChIP-chip procedures are that for ChIP-chip, probe preparation, hybridization, and scanning steps are needed instead of library preparation and sequencing (Fig. 1). Also, initial data processing and analysis (e.g., intensity normalization and peak calling) require different software because of the different format of raw data obtained by the two methods (sequence data from ChIP-seq vs. signal intensity data from ChIP-chip). Once information on DNA sequences bound by proteins of interest is obtained by either technique, downstream analyses can be done using the methods as described below. Since ChIP-seq has become a more widely used method for genome-wide analysis than ChIP-chip due to the reasons mentioned above and also for the rapidly decreasing sequencing cost, we will focus on ChIP-seq procedures in this chapter. However, for those who are interested in ChIP-chip, for example, using a specific chromosome tiling array, detailed procedures for the experiment and data analysis can be found in the previously published series [15–18].

2 Materials

2.1 Cell Culture

1. Growth medium (GM): Equal volume of Dulbecco's modified Eagle medium (DMEM) and Ham's F-10 Nutrient Mixture supplemented with 20 % fetal bovine serum, 1× penicillin–streptomycin, and 2.5 ng/ml basic fibroblast growth factor (human recombinant, EMD Millipore, Billerica, MA). Keep at 4 °C. Warm to 37 °C in a water bath when it is used.
2. Differentiation medium (DM): DMEM supplemented with 2 % horse serum and 1× penicillin–streptomycin. Keep at 4 °C. Warm to 37 °C in a water bath when it is used.
3. Phosphate-buffered saline (PBS): 136.9 mM NaCl, 2.7 mM KCl, 10.1 mM Na_2HPO_4, 1.8 mM KH_2PO_4. Keep at room temperature after autoclaving.
4. Collagen solution: Collagen from calf skin (Type I, 0.1 % solution in 0.1 M acetic acid, Sigma-Aldrich, St. Louis, MO) was diluted tenfold using filter-sterilized 0.1 M acetic acid. Keep at 4 °C.
5. Collagen-coated tissue culture dishes: Collagen solution (see above) was added to tissue culture dishes and spread to cover the entire area of the bottom surface of dishes. After incubation for 30 min at 37 °C, collagen solution was removed (*see* **Note 1**), and dishes were washed with PBS once. Collagen-coated dishes can be kept at room temperature after removing PBS.

2.2 ChIP

1. Imprint Chromatin Immunoprecipitation Kit (Sigma-Aldrich).
2. PBS supplemented with 1 mM $MgCl_2$ (PBS-Mg). Keep at room temperature.
3. 2 mM Disuccinimidyl glutarate (DSG): DSG (Thermo Fisher Scientific, Waltham, MA) was dissolved in dimethyl sulfoxide (DMSO) at 0.5 M, and then diluted to 2 mM by mixing into PBS/Mg rapidly. Prepare immediately before use.
4. 1 % Formaldehyde: 37 % formaldehyde solution was diluted to 1 % using PBS. Prepare immediately before use.
5. 1.25 M Glycine: 4.69 g of glycine was dissolved in 50 ml of double-distilled water (ddH_2O). Keep at 4 °C.
6. Bioruptor sonication device (Diagenode, Denville, NJ).
7. Antibodies (*see* **Note 2**): Anti-SOX6 antibody, Rabbit polyclonal, ChIP Grade (ab30455, Abcam, Cambridge, MA), Anti-RNA polymerase II CTD repeat YSPTSPS antibody [4H8], Mouse monoclonal, ChIP Grade (ab5408, Abcam), Normal mouse IgG (a component of Imprint Chromatin Immunoprecipitation Kit, Sigma-Aldrich).

2.3 qPCR

1. ABI Prism 7900 HT Sequence Detection System (Applied Biosystems, Foster City, CA).
2. Fast SYBR Green Master Mix (Applied Biosystems).
3. Primers: 5′-CCCCACCCCCTGGAACT and 5′-CCAGCTAGGAAACAATTGGAAGTG (for *Myh7* promoter), 5′-TCGGACCGGAGTGTTAGGAA and 5′-ACCCTGGAGTCTCAGCATCG (for an intergenic region located ~15 kb upstream of *Myh7*), 5′-AGCTGATGGCTGCAGGTTCT and 5′-GCCCCAGGCTATCTCATGTT (for *Gapdh* intron 2).

2.4 Library Preparation

1. NEBNext DNA Library Prep Reagent Set for Illumina (formerly called NEBNext DNA Sample Prep Reagent Set 1, New England Biolabs, Ipswich, MA).
2. Genomic DNA Sample Prep Oligo Only Kit (Illumina, San Diego, CA).
3. Bioruptor (Diagenode).
4. DNA Clean & Concentrator-5 (Zymo Research, Irvine, CA).
5. MinElute Gel Extraction Kit (Qiagen, Valencia, CA).
6. Thermal cycler.
7. Equipment for agarose gel electrophoresis.

2.5 Data Analysis

1. A computer installed with Linux operating system (OS): We used HP Compaq dc7900 Ultra-slim Desktop PC with E8500 Core2 Duo CPU and 4 GB memory installed with Ubuntu 10.04 LTS OS.
2. Bowtie: An ultrafast, memory-efficient short read aligner [19]. The program can be obtained at http://bowtie-bio.sourceforge.net/index.shtml.
3. ERANGE: A package of python scripts for analyzing high-throughput sequence data [20]. The package can be obtained at http://woldlab.caltech.edu/rnaseq/.
4. UCSC Genome Browser: A data visualization Web site [21] available at http://genome.ucsc.edu/.
5. SISSRs: A peak calling program [22, 23] which can be obtained at http://www.rajajothi.com/sissrs/.
6. ChIP-Seq Tool Set: Web-based software designed to analyze and compare ChIP-seq data [24] available at http://chipseq.genomecenter.ucdavis.edu/cgi-bin/chipseq.cgi/.
7. PeakAnalyzer: A peak annotation program [25] which can be obtained at http://www.bioinformatics.org/peakanalyzer/wiki/.
8. MEME: A motif discovery program [26] which can be obtained at http://meme.sdsc.edu/meme/.

3 Methods

All ChIP experiments were performed using in vitro-differentiated primary myotubes in which Sox6 expression is highly induced and a Sox6 target gene is significantly downregulated [13]. For ChIP-seq, we prepared two 15 cm dishes of myotube culture and performed ChIP separately, providing biological duplicates and therefore more reliable data by comparing the two independent data sets.

3.1 Cell Culture

1. All cell culture was performed at 37 °C in a 5 % CO_2-humidified incubator.
2. Primary myoblasts isolated from wild-type mouse E18.5 fetal limb muscle [13] were maintained in GM using collagen-coated dishes.
3. For all ChIP experiments, myoblasts were differentiated to myotubes by switching from GM to DM. For ChIP-seq, myoblasts were grown to nearly confluent in GM using 15 cm dishes. After rinsing with PBS once, DM was added and incubated for 48 h. Myotube formation was confirmed using a bright field microscope.

3.2 ChIP

1. Cells were washed with PBS-Mg once, and fixed with 2 mM DSG for 45 min at room temperature with gentle agitation (*see* **Note 3**). Cells were then washed with PBS-Mg twice, and fixed further using 1 % formaldehyde for 10 min at room temperature with gentle agitation.
2. Fixation was quenched by adding 1/10th volume of 1.25 M glycine. After incubating for 5 min at room temperature with gentle agitation, cells were washed three times with PBS and transferred to 1.5 ml microcentrifuge tubes by scraping. Cells were then centrifuged at ~180 × *g* for 5 min at 4 °C, and the supernatant was removed (*see* **Note 4**).
3. ChIP was performed according to the manufacturer's instructions (*see* **Note 5**). 1.4 ml of nuclei preparation buffer and 800 μl of shearing buffer were used for a 15 cm culture dish. Sonication conditions were as follows. Power setting: H (high). Cycling parameters: 15 s ON, 30 s OFF, 32 cycles (*see* **Note 6**). For ChIP-seq, one, two, and five wells were used for IgG, Pol II, and Sox6 antibodies, respectively (*see* **Note 7**). Purified DNA from the same antibody (eluted with 30 μl elution solution per well) was pooled, and stored at −20 °C until use.

3.3 qPCR

1. ChIP results were confirmed by qPCR using specific primers (*see* **Note 8**). qPCR was performed according to the manufacturer's instructions. PCR conditions used were as follows. Primer concentration: 0.2 μM; template DNA volume: 2 μl per 20 μl reaction; Taq polymerase activation: for 30 s at 95 °C; denaturation:

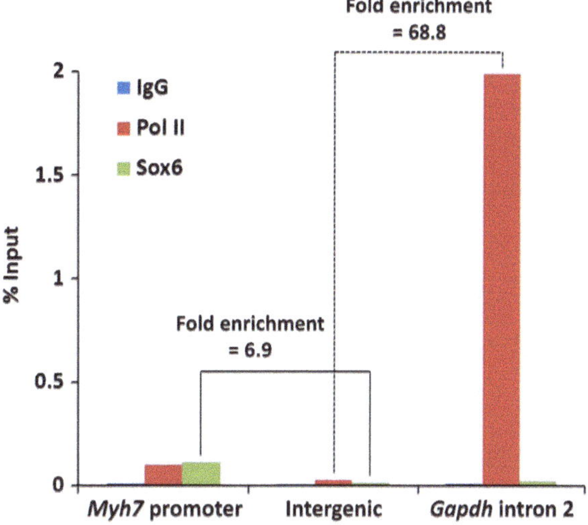

Fig. 2 An example of ChIP-qPCR results. Sox6 and Pol II enrichment measured in % input at the *Myh7* promoter (a known Sox6-binding site), an intergenic region (a negative control binding site), and the *Gapdh* intron 2 (a positive control binding site for Pol II) obtained from one of the two ChIP-seq samples are shown. Fold enrichment over the negative control region is also indicated

for 1 s at 95 °C; annealing/extension: for 20 s at 60 °C; cycles: 40. Single product was confirmed by dissociation curve analysis.

2. Results were normalized to input, and fold enrichment was calculated by normalizing to enrichment at a negative control region (intergenic). An example of the results obtained is shown in Fig. 2.

3.4 Library Preparation and Sequencing

Library preparation for sequencing with Illumina Genome Analyzer II was performed as previously described [27] with some modifications. Library preparation consists of the following five steps: end repair, A-tailing, adapter ligation, PCR amplification, and gel purification. We performed the PCR amplification step before the gel purification step to reduce the risk of complexity loss as described previously [27].

1. End repair was carried out by mixing the following constituents and incubating for 30 min at 20 °C in a thermal cycler:

ChIP sample + water	75 µl
10× Phosphorylation buffer	10 µl
dNTP solution mix (10 mM)	4 µl
T4 DNA polymerase	5 µl
DNA polymerase I, large (Klenow) fragment	1 µl
T4 Polynucleotide kinase	5 µl
Total	100 µl

End-repaired DNA was purified using DNA Clean & Concentrator-5 and eluted with 33 μl of preheated (at 50 °C) buffer EB (a component of MinElute Gel Extraction Kit).

2. A-tailing was performed by mixing the following constituents and incubating for 30 min at 37 °C in a thermal cycler:

DNA sample	32 μl
NEBuffer 2 (10×)	5 μl
dATP solution (1 mM)	10 μl
Klenow fragment (3′→5′ exo$^-$)	3 μl
Total	50 μl

A-tailed DNA was purified using DNA Clean & Concentrator-5 and eluted with 20 μl of preheated (at 50 °C) buffer EB.

3. Adapter ligation was conducted by mixing the following constituents and incubating for 15 min at room temperature:

DNA sample	19 μl
Quick ligation reaction buffer (2×)	25 μl
Genomic adapter Oligo mix	1 μl
Quick T4 DNA ligase	5 μl
Total	50 μl

Adapter-ligated DNA was purified using DNA Clean & Concentrator-5 and eluted with 25 μl of preheated (at 50 °C) buffer EB.

4. PCR amplification was performed by mixing the following constituents and running the following program:
(Reaction mix):

DNA sample	24 μl
Phusion high-fidelity PCR master mix (2×)	25 μl
PCR primer 1.1	0.5 μl
PCR primer 2.1	0.5 μl
Total	50 μl

(PCR program):

Step 1: 98 °C, 30 s.

Step 2: 98 °C, 10 s.

Step 3: 65 °C, 30 s.

Step 4: 72 °C, 30 s.

Step 5: Go to **step 2**, repeat 18 times.

Step 6: 72 °C, 5 min.

Step 7: 4 °C.

5. Gel purification was done by using 2 % agarose gel. Gels containing approx. 300–600 bp fragments were excised (*see* **Note 9**) using a clean razor blade, and DNA was extracted using MinElute Gel Extraction Kit. Gel slices were dissolved at room temperature as described previously [28], and DNA was eluted with 15 μl of preheated (at 50 °C) buffer EB. Extracted DNA was further purified using DNA Clean & Concentrator-5.
6. Quality and quantity of the library were checked with Agilent Bioanalyzer 2100 (*see* **Note 10**), and sequenced for 40 cycles using a single read mode on an Illumina Genome Analyzer II at the DNA Technologies Core Facility at UC Davis (http://dnatech.genomecenter.ucdavis.edu/).

3.5 Data Analysis

Although data analysis steps can be different depending on the purpose of each study, there are common basic steps including alignment of short reads to the reference genome, visualization, peak calling, and annotation of peaks. Here we show an example procedure of these steps together with additional analysis steps including finding common peaks from two data sets and motif discovery. Because some programs are found to be updated since our last analysis, it is advisable to check updated information and use the latest version when you follow the procedures below.

1. ChIP-seq and input sample reads (which were all in the Illumina native text file format) were aligned using Bowtie (version 0.12.5) to the mouse NCBI Build 37 genome assembly. An example of a command line can be found at ERANGE Web site (*see* the section 6 at http://woldlab.caltech.edu/erange/README.build-rds).
2. Aligned sequence data were visualized at UCSC Genome Browser using BED format files (*see* **Note 11**). BED files were created from Bowtie output files via RDS format files using ERANGE (version 3.2). Only uniquely mapped reads were included in BED files by specifying the "-nomulti" option to minimize complexity. An example of visualized ChIP-seq data is shown in Fig. 3.
3. Peak calling was performed with SISSRs (version 1.4) using the following parameters: -i (input file): aligned read data in BED format (see above), -s (genome size): 2,716,965,481 bp, -F (average length of DNA fragments): 450 bp, -b (background file): input DNA data file of each ChIP experiment (*see* **Note 12**).

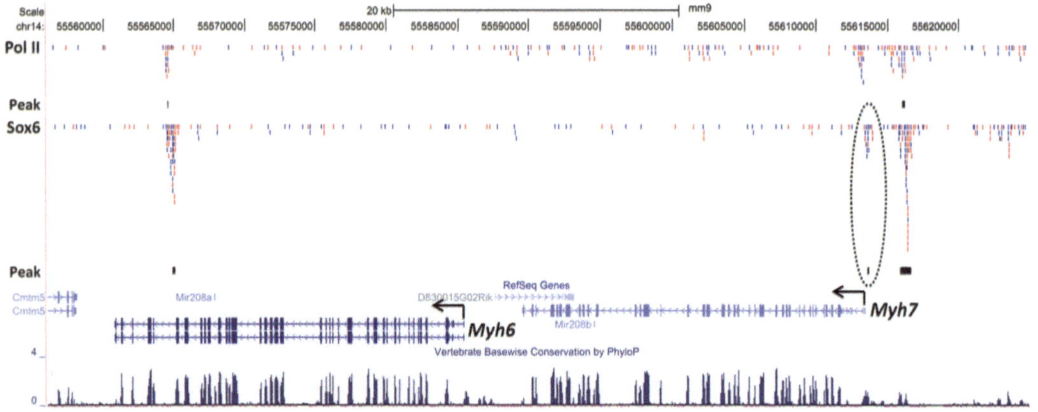

Fig. 3 Sox6- and Pol II-binding events at the *Myh6*/*Myh7* loci revealed by ChIP-seq. One of the two ChIP-seq data sets (aligned read data in BED format) is shown together with peaks identified by SISSRs (*solid black boxes*, in GFF format). The chromosomal position of mouse NCBI37/mm9 assembly (*top*), sequence conservation (*bottom*), and transcription start sites of *Myh6* and *Myh7* (*arrow*) are also shown. A *broken circle* indicates a known Sox6-binding site on the Myh7 promoter, confirming validity of the ChIP-seq data. *Myh6*: myosin, heavy polypeptide 6, cardiac muscle, alpha, *Myh7*: myosin, heavy polypeptide 7, cardiac muscle, beta

4. Common peaks from the two Sox6 ChIP-seq data sets were identified at the ChIP-seq Tool Set Web site. Input files in GFF format were created from SISSR output files by modifying on Microsoft Excel according to the file format description on the UCSC Genome Browser Web site (http://genome.ucsc.edu/FAQ/FAQformat.html). In our analysis, corresponding peaks within 50 bp were considered as overlapping. Results (in GFF format) can be visualized at UCSC Genome Browser (*see* Fig. 3) (*see* **Note 13**).
5. Peak annotation was conducted with PeakAnalyzer using the default mouse mm9 annotation file.
6. Motif discovery was performed using MEME (version 4.5.0) with default parameters (*see* **Notes 14** and **15**). Input sequences of Sox6 peaks in FASTA format were obtained using Table Browser (http://genome.ucsc.edu/cgi-bin/hgTables/) by specifying the common Sox6 peak track as an input track, and "sequence" as an output format. An example of MEME results is shown in Fig. 4.

4 Notes

1. Removed collagen solution can be kept at 4 °C and reused at least once.
2. Use ChIP-grade or ChIP-validated antibody if available commercially. If not, you need to identify an antibody suitable for ChIP by testing its specificity and enrichment efficiency.

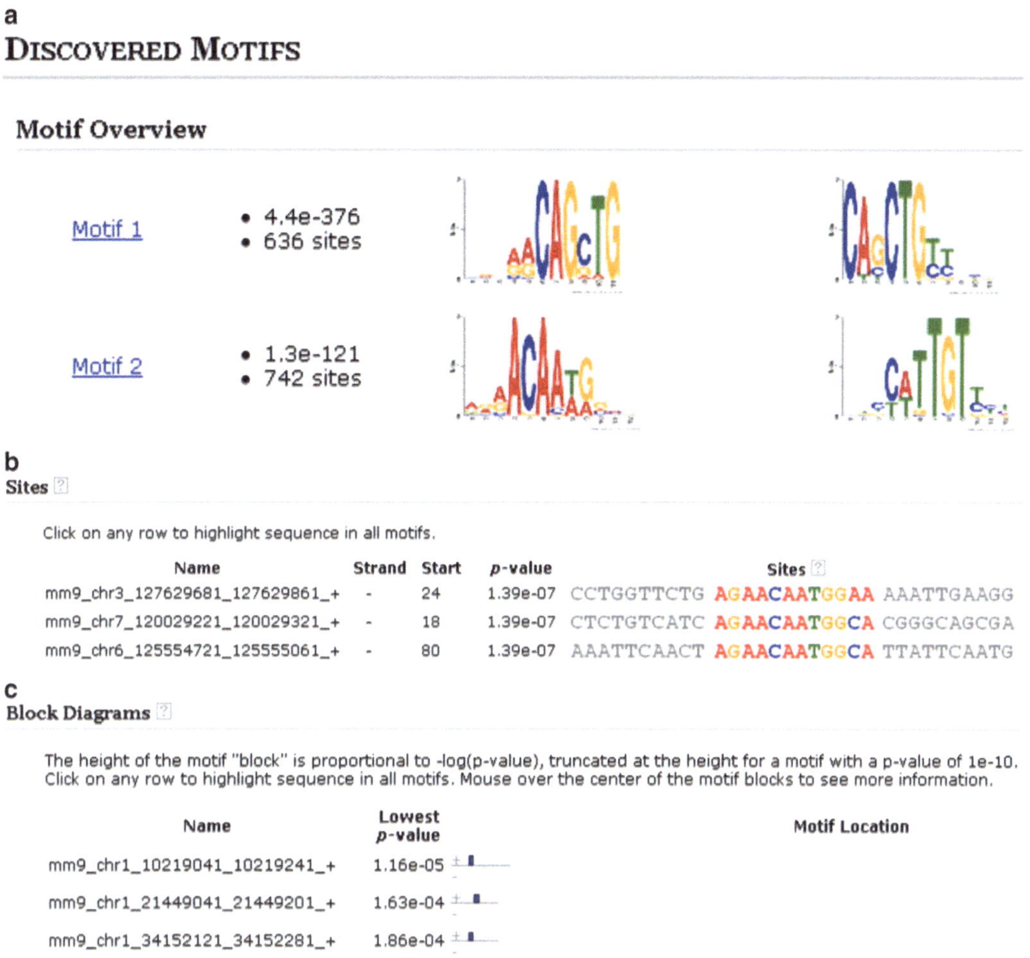

a

DISCOVERED MOTIFS

Motif Overview

Motif 1
- 4.4e-376
- 636 sites

Motif 2
- 1.3e-121
- 742 sites

b

Sites

Click on any row to highlight sequence in all motifs.

Name	Strand	Start	*p*-value	Sites		
mm9_chr3_127629681_127629861_+	-	24	1.39e-07	CCTGGTTCTG	AGAACAATGGAA	AAATTGAAGG
mm9_chr7_120029221_120029321_+	-	18	1.39e-07	CTCTGTCATC	AGAACAATGGCA	CGGGCAGCGA
mm9_chr6_125554721_125555061_+	-	80	1.39e-07	AAATTCAACT	AGAACAATGGCA	TTATTCAATG

c

Block Diagrams

The height of the motif "block" is proportional to -log(p-value), truncated at the height for a motif with a p-value of 1e-10.
Click on any row to highlight sequence in all motifs. Mouse over the center of the motif blocks to see more information.

Name	Lowest *p*-value	Motif Location
mm9_chr1_10219041_10219241_+	1.16e-05	
mm9_chr1_21449041_21449201_+	1.63e-04	
mm9_chr1_34152121_34152281_+	1.86e-04	

Fig. 4 An example of MEME results. A subset of results obtained with default parameters are shown when sequences of Sox6 common peaks (1,066 in total) were used as an input. (**a**) An example of "Motif Overview" screen. *Top* two motifs discovered (motif 1: *E-box*, motif 2: Sox consensus motif) are shown here. (**b**) An example of "Sites" screen of motif 2. Sequences of *top* three results are shown here. (**c**) An example of "Block Diagrams" screen of motif 2. Motif locations in the first three input sequences are shown here

3. Fixation conditions must be optimized empirically. Most ChIP protocols use only formaldehyde for fixation, but an additional cross-linking reagent (DSG) [29] was found to be essential in our case.
4. Fixed cells should be kept on ice until use. If ChIP is not conducted immediately, pellets of fixed cells can be flash-frozen in liquid nitrogen and stored at −80 °C. (Storing frozen fixed cells does not appear to affect results. However, freezing cell lysate is not recommended.)
5. Any successful ChIP protocol or kit can be used.
6. Sonication conditions must be optimized empirically. We used ChIP-qPCR results in addition to DNA size as indices for optimization.

7. With this number of wells, we obtained 3 and 1.5 million of uniquely mapped reads out of ~20 million of total reads from two independent Sox6 ChIP-seq experiments [8]. We speculate that this relatively small number of uniquely mapped reads is due to the low amount of immunoprecipitated DNA. Therefore, we anticipate that more uniquely mapped reads could be obtained if a larger number of wells were used for a single ChIP-seq experiment.
8. Prior to library construction, ChIP results must be confirmed by qPCR using appropriate negative and positive control primers. As for negative controls, there should be no or very low enrichment with IgG antibody for any target sites. Also, there should be no or very low enrichment for a nontarget site (e.g., an intergenic region) with any antibodies used. An example is presented in Fig. 2.
9. The desired size can vary depending on your samples. In our case, 300–600 bp fragments gave more and larger peaks than 200–300 bp fragments.
10. It is recommended to perform qPCR using the same primer sets as ChIP-qPCR (*see* Subheading 3.3) to check the library quality.
11. Other file formats can be used for visualization of ChIP-seq data. For details, refer to the UCSC Genome Browser Web site (http://genome.ucsc.edu/FAQ/FAQformat.html).
12. It is strongly recommended to test several peak calling programs because different programs can produce different results (e.g., different peak numbers, peak positions, peak width) depending on the quality and quantity of sequence data used. We chose SISSRs among several programs tested based on the observation that the results obtained with this program matched best with visually identified peaks on UCSC Genome Browser.
13. The same approach can be used when comparing binding sites of two different transcription factors [8].
14. It is recommended to test different settings for motif distribution (which can be specified using the "-mod" option) because distinctive motifs can be found with different settings as shown previously [8].
15. The MEME program can be run on the MEME Web site (http://meme.sdsc.edu/meme/cgi-bin/meme.cgi) as well. However, there is a limit of character numbers in input sequences (up to 60,000 characters). Therefore, if your sequence data has more than 60,000 characters, you need to either trim the data or run an installed program on your computer with the "-maxsize" option (e.g., 1,000,000).

Acknowledgments

We thank Mr. Adam Jenkins for helpful discussions and Dr. Charles Nicolet at the DNA Technologies and Expression Analysis Core Facilities of the UC Davis Genome Center for assistance with the ChIP-seq experiments. This work was supported by Expression Analysis Core Seed Grant, Muscular Dystrophy Association (MDA 4135), and the National Institutes of Health (AR055209) (to N.H.).

References

1. Solomon MJ, Larsen PL, Varshavsky A (1988) Mapping protein-DNA interactions in vivo with formaldehyde: evidence that histone H4 is retained on a highly transcribed gene. Cell 53:937–947
2. Ren B, Robert F, Wyrick JJ, Aparicio O, Jennings EG, Simon I et al (2000) Genome-wide location and function of DNA binding proteins. Science 290:2306–2309
3. Johnson DS, Mortazavi A, Myers RM, Wold B (2007) Genome-wide mapping of in vivo protein-DNA interactions. Science 316: 1497–1502
4. Hurd PJ, Nelson CJ (2009) Advantages of next-generation sequencing versus the micro-array in epigenetic research. Brief Funct Genomic Proteomic 8:174–183
5. Park PJ (2009) ChIP-seq: advantages and challenges of a maturing technology. Nat Rev Genet 10:669–680
6. Chen Y, Negre N, Li Q, Mieczkowska JO, Slattery M, Liu T et al (2012) Systematic evaluation of factors influencing ChIP-seq fidelity. Nat Methods 9:609–614
7. Rhee HS, Pugh BF (2011) Comprehensive genome-wide protein-DNA interactions detected at single-nucleotide resolution. Cell 147:1408–1419
8. An CI, Dong Y, Hagiwara N (2011) Genome-wide mapping of Sox6 binding sites in skeletal muscle reveals both direct and indirect regulation of muscle terminal differentiation by Sox6. BMC Dev Biol 11:59
9. Bowles J, Schepers G, Koopman P (2000) Phylogeny of the SOX family of developmental transcription factors based on sequence and structural indicators. Dev Biol 227:239–255
10. Hagiwara N (2011) Sox6, jack of all trades: a versatile regulatory protein in vertebrate development. Dev Dyn 240:1311–1321
11. Kamachi Y, Uchikawa M, Kondoh H (2000) Pairing SOX off: with partners in the regulation of embryonic development. Trends Genet 16:182–187
12. Wegner M (1999) From head to toes: the multiple facets of Sox proteins. Nucleic Acids Res 27:1409–1420
13. Hagiwara N, Yeh M, Liu A (2007) Sox6 is required for normal fiber type differentiation of fetal skeletal muscle in mice. Dev Dyn 236:2062–2076
14. Hagiwara N, Ma B, Ly A (2005) Slow and fast fiber isoform gene expression is systematically altered in skeletal muscle of the Sox6 mutant, p100H. Dev Dyn 234:301–311
15. Cauchy P, Benoukraf T, Ferrier P (2011) Processing ChIP-chip data: from the scanner to the browser. Methods Mol Biol 719: 251–268
16. Gobel U, Reimer J, Turck F (2010) Genome-wide mapping of protein-DNA interaction by chromatin immunoprecipitation and DNA microarray hybridization (ChIP-chip). Part B: ChIP-chip data analysis. Methods Mol Biol 631:161–184
17. Reimer JJ, Turck F (2010) Genome-wide mapping of protein-DNA interaction by chromatin immunoprecipitation and DNA microarray hybridization (ChIP-chip). Part A: ChIP-chip molecular methods. Methods Mol Biol 631:139–160
18. Spiro S (2012) Genome-wide mapping of the binding sites of proteins that interact with DNA. Methods Mol Biol 881:137–156
19. Langmead B, Trapnell C, Pop M, Salzberg SL (2009) Ultrafast and memory-efficient alignment of short DNA sequences to the human genome. Genome Biol 10:R25
20. Mortazavi A, Williams BA, McCue K, Schaeffer L, Wold B (2008) Mapping and quantifying mammalian transcriptomes by RNA-Seq. Nat Methods 5:621–628
21. Kent WJ, Sugnet CW, Furey TS, Roskin KM, Pringle TH, Zahler AM, Haussler D (2002)

The human genome browser at UCSC. Genome Res 12:996–1006

22. Jothi R, Cuddapah S, Barski A, Cui K, Zhao K (2008) Genome-wide identification of in vivo protein-DNA binding sites from ChIP-Seq data. Nucleic Acids Res 36:5221–5231
23. Narlikar L, Jothi R (2012) ChIP-Seq data analysis: identification of protein-DNA binding sites with SISSRs peak-finder. Methods Mol Biol 802:305–322
24. Blahnik KR, Dou L, O'Geen H, McPhillips T, Xu X, Cao AR et al (2010) Sole-Search: an integrated analysis program for peak detection and functional annotation using ChIP-seq data. Nucleic Acids Res 38:e13
25. Salmon-Divon M, Dvinge H, Tammoja K, Bertone P et al (2010) PeakAnalyzer: genome-wide annotation of chromatin binding and modification loci. BMC Bioinformatics 11:415
26. Bailey TL, Elkan C (1994) Fitting a mixture model by expectation maximization to discover motifs in biopolymers. Proc Int Conf Intell Syst Mol Biol 2:28–36
27. Schmidt D, Wilson MD, Spyrou C, Brown GD, Hadfield J, Odom DT et al (2009) ChIP-seq: using high-throughput sequencing to discover protein-DNA interactions. Methods 48:240–248
28. Quail MA, Kozarewa I, Smith F, Scally A, Stephens PJ, Durbin R et al (2008) A large genome center's improvements to the Illumina sequencing system. Nat Methods 5: 1005–1010
29. Nowak DE, Tian B, Brasier AR (2005) Two-step cross-linking method for identification of NF-kappaB gene network by chromatin immunoprecipitation. Biotechniques 39: 715–725

Chapter 5

Analysis of Allele-Specific Gene Expression Using a Target-Oriented Tiling Microarray Assay

Huiyong Zhang, Xing-Wang Deng, and Lei Li

Abstract

Advances in molecular and computational biology in recent years have led to the development or the improvement of methods for analyzing global gene expression. In most of these efforts, it is assumed that alleles of different origins contribute equally to the transcript pool. However, accumulating evidence suggests that many genes are not equally expressed from the paternal and maternal chromosomes. In addition to imprinting, the phenomenon of imbalanced allelic expression is widespread in heterozygous individuals. To distinguish transcript pools derived from different alleles present in the same organism, a number of methods have been developed. Here, we describe an oligonucleotide tiling microarray-based assay for analyzing allele-specific gene expression. Specifically targeting single-nucleotide polymorphisms, this two-step assay offers a high-throughput and multiplex method for detecting and quantifying unequal allelic expression that is readily applicable to many experimental systems.

Key words Gene expression, Tiling microarray, SNP, Allele, Hybrid, Rice

1 Introduction

Analysis of global gene expression provides valuable insight into biological processes and developmental programs. When more than one set of alleles are present in the same organism, total transcript output from a locus is determined by allele-specific variation (*cis* effect), variation in other regulatory genes (*trans* effect), as well as environmental influences. There have been numerous reports that many genes are not equally expressed from the paternal and maternal alleles [1–3]. At the extreme are the imprinted genes exclusively transcribed from the non-silenced allele. In human, it is now clear that many genes exhibit imbalanced allele expression (IAE) in somatic cells of heterozygous individuals [4–7]. Further, mono-allelic expression patterns are increasingly being associated with loss of heterozygosity in tumor cells [8]. In plants, investigations of the transcript levels also indicate that IAE is potentially prevalent in heterotic F_1 hybrids [3, 9, 10].

Tin-Lap Lee and Alfred Chun Shui Luk (eds.), *Tiling Arrays: Methods and Protocols*, Methods in Molecular Biology, vol. 1067,
DOI 10.1007/978-1-62703-607-8_5, © Springer Science+Business Media New York 2013

Furthermore, numerous studies in polyploid plant species on subsets of the genome indicate that unequal expression of the homologous alleles is quite common [11–13]. Therefore, experimental capacity to discern allele-specific expression at the genome scale is much desired.

A vast majority of experimental efforts to distinguish the transcript of one allele from its highly similar counterpart have been directed toward single-nucleotide polymorphisms (SNPs), the most abundant form of DNA polymorphism. Experimental procedures based on chemical and physical principles to examine unequal transcription from SNP-defined alleles were successfully developed [5, 9, 14–16]. In addition, high-throughput approaches including microarray analysis and whole-transcriptome RNA sequencing were employed to detect and quantify allele-specific transcripts. Although RNA sequencing is a powerful method for detecting allele-specific expression, new statistical methods need to be developed to capitalize on the wealth of information [17] and to avoid false-positives due to technical bias of the experimental approach [18].

In contrast to RNA sequencing, microarray-based applications, which include mini-sequencing on microarray [19] and allele-specific microarrays [7, 20, 21], are target oriented and suitable for monitoring a predefined set of genes of interest. These applications offer a high-throughput and multiplex method for quantifying unequal allelic expression that is affordable to most research laboratories. These microarrays typically involve pairs of probes with perfect match to one of the two alleles, based on the idea that a sequence mismatch may significantly disrupt hybridization and attenuate a probe's signal to the non-matching target. However, two key technical issues need to be fully addressed in developing genome-scale allele-specific microarrays. First, the designed probes should offer sufficient sensitivity and specificity in discriminating the transcripts derived from different alleles. Second, the relative hybridization signal from the allele-specific probes should respond linearly to changes in the allele-specific transcript level. In these regards, the tiling microarray technology, which involves the representation of a genomic region with progressive oligonucleotide probes [22–28], offers a versatile and powerful platform for the design of allele-specific microarrays.

He et al. [29] developed a tiling microarray to quantify allele-specific expression from 518 SNPs between two rice subspecies *indica* and *japonica*, which involved 25 blocks of 25-mer probes for each SNP. Each probe block consisted of two allele-specific probes matching perfectly to the *indica* and *japonica* alleles and two control probes according to the two remaining possible nucleotides. A series of samples, which contains the *japonica* and *indica-derived* cDNA templates mixed at different proportions, were hybridized to the tiling arrays. For each probe block, the

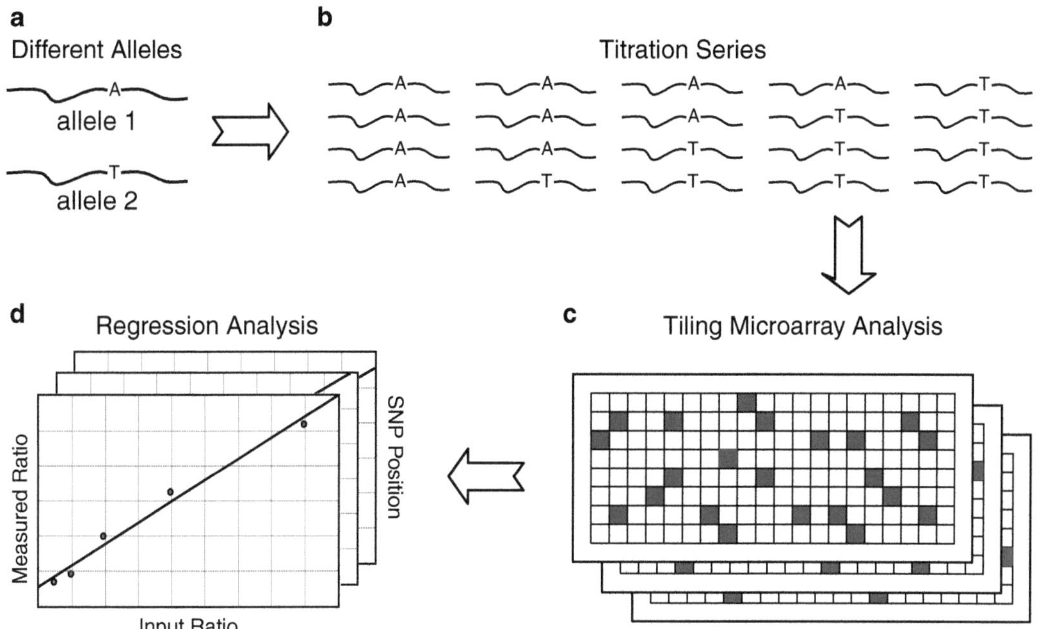

Fig. 1 Workflow of the tiling microarray approach to screen for diagnostic probes for distinguishing allele-specific transcripts. (**a**) The oligonucleotide tiling microarray was designed to target SNP-bearing alleles by including all possible probes interrogating each SNP. (**b**) RNA from individuals harboring SNPs is reverse-transcribed and mixed with altering proportions to make a titration series. (**c**) The tiling microarray is hybridized to the titration series in which the input ratios of the two alleles at the transcript level are known. (**d**) The probes interrogating each SNP are screened based on a regression model to identify those that show strong linearity between the measured and the input ratio of allele-specific transcripts over the titration series

ratios of *indica*- and *japonica*-specific transcript were correlated to the input ratios using a regression model. Based on a cutoff for significant linearity, over half of the SNPs were found to be able to accurately quantify allele-specific transcripts [29]. Here we describe in detail this method, which should be readily applicable to species with known DNA polymorphism information (summarized in Fig. 1).

2 Materials

2.1 Biological Materials

Biological samples should be prepared from two inbred lines with SNPs and the hybrids or heterozygotes. For the exemplary rice work, *Oryza sativa* L. ssp *indica* (93-11), *Oryza sativa* L. ssp *japonica* (Nipponbare), and their reciprocal hybrids were used. Seedlings were grown in soil in environmentally controlled growth chambers at 28 °C for 4 weeks until the four-leaf stage and harvested for RNA isolation.

2.2 RNA Isolation and Reverse Transcription

1. TRIzol reagent (Invitrogen).
2. Oligotex® mRNA Mini Kit (Qiagen).
3. Phenol:chloroform mixture (equal volume), at pH 4.5 (for RNA) or 8.0 (for DNA).
4. Nanodrop spectrophotometer (Thermo Scientific).
5. Random Hexamer primer (Roche).
6. SuperScript II reverse transcriptase kit (Invitrogen).

2.3 cDNA Sample Labeling, Microarray Hybridization, and Scanning

1. Amino-allyl-dUTP (aa-dUTP; Sigma).
2. Monofunctional Cy3 Dye (GE Healthcare).
3. Microcon YM-30 centrifugal filter column (Millipore).
4. QIAquick PCR Purification Kit (Qiagen).
5. Prehybridization buffer: 5×SSC, 0.1 % SDS, 1 % BSA.
6. Hybridization buffer: 5× SSC, 0.1 % SDS, 0.5 mg/ml BSA, 0.5 mg/ml salmon sperm DNA.
7. Wash buffers: (a) 2× SSC, 0.1 % SDS; (b) 0.2× SSC, 0.1 % SDS; (c) 0.2× SSC.
8. GenePix 4200A scanner (Axon).

2.4 Quantitative Real-Time PCR Analysis

1. SYBR® Green Master Mix (Applied Biosystems).
2. qPCR 96-well plates and seals (Applied Biosystems).
3. ABI 7500-Fast Real-Time PCR systems (Applied Biosystems).

3 Methods

3.1 SNP Selection

Initial SNP information can be queried and downloaded from the NCBI dbSNP database (http://www.ncbi.nlm.nih.gov/snp) or other appropriate sequence repositories. To facilitate the selection of high-quality and informative SNPs that generate sufficient discrimination in attenuated hybridization intensities resulting from single base pair mismatch, we suggest the following criteria (*see* **Note 1**):

1. The SNP candidates should be single polymorphic point per codon and locate in coding regions supported by full-length cDNAs or expressed sequence tags.
2. Synonymous and non-synonymous SNP candidates should occupy a comparable frequency.
3. The polymorphic nucleotides should have a high sequencing quality score [30].
4. For two adjacent SNP sites within the same gene to be included, they should be separated by at least 50 bp.

5. The 25-mer probes covering the SNP site should have a GC content range of 40–60 %, and a melting temperature in the range of 70 ± 5 °C.
6. Similarity search should be performed by blasting the 50 bp sequences surrounding the candidate SNPs against the appropriate genome to remove highly repetitive sequences.

3.2 Design of SNP-Based Tiling Microarray

The tiling strategy was previously described [29]. In this design, 25-mer oligonucleotide probes are used. To assess the positional effect of the polymorphism within the probe, each SNP is covered by 25 probe sets with the polymorphic site being placed at all 25 possible positions within the probe. Thus all possible 25-mer probes spanning the polymorphic site for each SNP are included in the tiling design. For each position, four probes are used that consist of a unit (a probe block) for measuring allelic output. Two of these match perfectly to one of the two alleles (e.g., *indica* and *japonica* rice) in question. The other two probes, called control probes, contain one of the two remaining nucleotides at the SNP site, which are used as negative control to estimate the baseline of background signals and to correct false detections. In addition, four allele-neutral probes are selected from the flanking neighborhood of each SNP site to control for overall transcript level from a given locus. Thus, a total of 104 ($1 \times 25 \times 4 + 1 \times 4$) probes will be designed to tile each SNP.

In theory, the probes can be synthesized on any microarray platform. We recommend the use of microarray produced on the NimbleGen platform based on our own experience [23, 29]. Probe synthesis for NimbleGen microarrays involves progressive photodeprotection of sites at oligonucleotide extension on the array surface, followed by DNA synthesis chemistries. Because photodeprotection is achieved by steering the UV light onto the array surface using a matrix of digitally controlled micromirrors, the microarrays are truly custom-made and offer greater flexibility than other microarray technologies.

3.3 Extraction of Total RNA

RNA is isolated using TRIzol reagent according to the manufacturer's instructions.

1. Briefly incubate the homogenized samples (50–100 mg per 1 ml TRIzol) for 15 min at room temperature to permit complete dissociation of the nucleoprotein complex.
2. Add 0.2 ml of chloroform per 1 ml of TRIzol reagent, and vortex vigorously for 15 s.
3. Incubate at room temperature for 5 min, and centrifuge the sample at 12,000 × *g* for 15 min at 4 °C.

4. Remove the aqueous phase of the sample into a new tube by pipetting. Avoid drawing any of the interphase or organic layer into the pipette when removing the aqueous phase.
5. Add 0.5 ml of 100 % isopropanol to the aqueous phase, per 1 ml of TRIzol reagent used for homogenization. Incubate at room temperature for 10 min, and centrifuge at 12,000 × *g* for 10 min at 4 °C.
6. Remove the supernatant and wash the RNA pellet with 75 % ethanol.
7. Air-dry the RNA pellet, and dissolve in nuclease-free water.

3.4 Removal of DNA and Quantification of RNA Concentration

1. Set up the DNase digestion reaction (1 U of DNase I per 1 μg of total RNA): RNA sample, 5 μl of 10× DNase I buffer, DNase I enzyme, and nuclease-free water to a final volume of 50 μl. Incubate in 37 °C for 30 min.
2. Add 5 μl of RQ1 DNase stop solution to terminate the reaction, and incubate at 65 °C for 10 min to inactivate the DNase.
3. Add equal volume of phenol:chloroform mixture to the reaction system, and vortex for 15 s.
4. Centrifuge at 12,000 × *g* for 3 min at room temperature, and transfer the supernatant to a new 1.5 ml DNase/RNase-free tube.
5. Repeat **steps 3** and **4** using equal volume of chloroform to remove residual phenol from the RNA sample. Add 1/10 volume of 3 M NaAC (pH 5.2) and equal volume of isopropanol to precipitate RNA.
6. Measure the concentration and the quality of the RNA in the NanoDrop spectrophotometer using the program for RNA (*see* **Note 2**).

3.5 Purification of Poly A$^+$ RNA

Isolation of Poly A$^+$ RNA from total RNA is performed using Oligotex mRNA spin-column according to the manufacturer's instruction. See the user guide for more details with various amount of starting total RNA. In our experiment, we start from 500 μg of total RNA.

1. Pipet total RNA into an RNase-free 1.5 ml microcentrifuge tube, and adjust the volume with nuclease-free water to 500 μl.
2. Add 500 μl of buffer OBB and 30 μl of Oligotex suspension. Mix the contents thoroughly by pipetting.
3. Incubate the sample at 70 °C for 3 min, and keep the sample at room temperature for 10 min.
4. Centrifuge the sample complex at 16,000 × *g* for 2 min, and carefully remove the supernatant by pipetting.

5. Resuspend the Oligotex:mRNA pellet in 400 μl of buffer OW2 by vortexing, and pipet onto a small spin column. Centrifuge at 16,000 × *g* for 1 min.
6. Transfer the column to a new RNase-free 1.5 ml microcentrifuge tube, and apply 400 μl of buffer OW2 to the column. Centrifuge at 16,000 × *g* for 1 min and discard the flow-through.
7. Transfer spin column to a new RNase-free 1.5 ml microcentrifuge tube. Pipet 50 μl hot (70 °C) buffer OEB onto the column, pipet up and down to resuspend the resin, and centrifuge at 16,000 × *g* for 1 min.
8. Quantitate the purified mRNA by a NanoDrop spectrophotometer.

3.6 cDNA Synthesis

1. In a 0.2 ml tube, add 200 ng of Poly A$^+$ RNA, 1 μl of oligo(dT)$_{18}$ primer, 1 μl of 10 mM dNTP, and DNase/RNase-free distilled water to a final volume of 12 μl. Program a hot start in the thermal cycler: 5 min at 65 °C followed by 3 min at 4 °C and load tubes. When the PCR block has been at 4 °C for 2 min, take the tubes from the block and centrifuge shortly to collect the entire samples in the bottom of the tube.
2. Add 4 μl of 5× first-strand buffer, 2 μl of 0.1 M DTT, 1 μl of RNase inhibitor, and 1 μl of the superscript enzyme. A master mix should be prepared to eliminate variations in the composition of the reaction between tubes. The cDNA synthesis is performed for 1 h at 42 °C, followed by 15 min at 75 °C, and 4 °C forever.
3. Dilute the cDNA samples 1:5 in DNase/RNase-free distilled water as templates for quantitative real-time PCR analysis. These cDNA samples can be stored at 4 °C for a short period and −20 °C for a long term.

3.7 Sample Labeling

1. Incorporate aa-dUTP into cDNA sample. Add the following components in a 0.2 ml tube for a volume of 20 μl: 10 μg of purified mRNA sample, 3 μl of oligo(dT)$_{18}$ (1 μg/μl), 2 μl of hexamer (2.5 μg/μl), DNase/RNase-free water to a total volume of 20 μl.
2. Mix well, incubate at 68 °C for 4 min, and then place on ice.
3. Prepare the following mixture, and add to each denatured RNA sample 8 μl of 5× first-strand buffer; 4 μl of 0.1 M DTT; 0.6 μl of the mix of dATP, dCTP, and dGTP (each 33.3 mM); 0.4 μl of 25 mM dTTP; 4 μl of 2 mM aa-dUTP; 1 μl of RNsin (10 U/μl); and 2 μl of SuperScript II reverse transcriptase.
4. Incubate at 42 °C for 1 h, add another 2 μl of SuperScript II reverse transcriptase, and continue to incubate at 42 °C for 2 h.

5. Heat at 94 °C for 3 min to inactivate reverse transcriptase.
6. Add 2 μl of RNase (0.5 μg/μl) and incubate at 37 °C for 15 min to remove RNA.
7. Add equal volume of phenol:chloroform mixture to each cDNA sample, and vortex.
8. Centrifuge at 16,000 × *g* for 5 min at room temperature.
9. Transfer the supernatant to a micro YM-30 column, and add 500 μl of nuclease-free water.
10. Centrifuge at 16,000 × *g* for about 15 min at room temperature until the volume left no more than 10 μl.
11. Add 500 μl of 0.1 M $NaHCO_3$ (pH 9.0), and perform the centrifuge until the volume left no more than 10 μl (*see* **Note 3**).
12. Invert the column in a 1.5 ml collection tube, and centrifuge at 16,000 × *g* for 1 min to collect the cDNA probe.
13. Add 1 μl of monofunctional CY3 (one kit in 20 μl DMSO) to each purified cDNA probe, mix well, and incubate in the dark at room temperature for 60–90 min (*see* **Note 4**).
14. Add 1 μl of ethanolamine to stop the labeling reaction.
15. Transfer each labeled cDNA probe to a separate micro YM-30 column, and add 500 μl of nuclease-free water. Centrifuge at 16,000 × *g* for about 15 min at room temperature until the volume left no more than 10 μl.
16. Add another 500 μl of nuclease-free water, and repeat the centrifuge above.
17. Collect the labeled cDNA probe in a new 1.5 ml tube, and quantitate each sample using the NanoDrop spectrophotometer (*see* **Note 5**).

3.8 Microarray Hybridization, Wash, and Scan

1. Prehybridize microarray slides with prehybridization buffer (5× SSC, 0.1 % SDS, 1 % BSA) at 42 °C for 1 h.
2. Wash the slides sequentially with water for 5 min, 70 % ethanol for 2 min, and 100 % ethanol for 2 min at room temperature. Air-dry the slides for hybridization use.
3. A total of 2 μg of Cy3 dye-labeled cDNA probe was used for each microarray slide. Base on different hybridization strategies, prepare the following probe mixtures (*see* **Note 6**):
   (a) For the titration series hybridization, the dye-labeled probes from the two inbreed lines (e.g., *indica* and *japonica*) were mixed with the following proportions: *indica*:*japonica* = 1:0; *indica*:*japonica* = 1:4; *indica*:*japonica* = 1:2; *indica*:*japonica* = 1:1; *indica*:*japonica* = 2:1; *indica*:*japonica* = 4:1; and *indica*:*japonica* = 0:1.
   (b) For testing hybrids or heterozygous individuals (e.g., reciprocal rice hybrids), three Cy3 dye-labeled probe from

indica × *japonica* or *japonica* × *indica* hybrids were independently prepared for triplicate hybridization.

4. Add each above probe mix to hybridization buffer (5× SSC, 0.1 % SDS, 0.5 mg/ml salmon sperm DNA, 0.5 mg/ml BSA), and hybridize with the pretreated slides at 42 °C for 16–20 h.
5. After hybridization, wash the slides sequentially with 2× SSC/0.1 % SDS, 0.2× SSC/0.1 % SDS, 0.2× SSC for 10 min each at room temperature.
6. Spin at 1,000 × rpm for 3 min in a 50 ml conical tube to remove excess wash buffer, and allow to air-dry completely.
7. Scan the dried slides under Cy3 channel at 5 μm resolution with a GenePix 4200A scanner, and save independent TIFF images for further analysis.

3.9 Microarray Data Analysis

3.9.1 Identification of Diagnostic Probe Blocks

Data from the titration series experiments is analyzed to identify diagnostic probe blocks that generate hybridization data quantitatively reflecting the relative abundance of allele-specific transcript in the inputs.

1. The obtained hybridization intensity for all probes is first log2-transformed and quantile-normalized across the several arrays.
2. For each block, the average value of the two control probes is used as background. The intensity for the allele-specific probe (*indica* or *japonica*) is calculated by subtracting the background value from the observed value. A baseline value of zero is used when negative values are encountered.
3. The relative intensity (after background correction) of the two allele-specific probes is used to determine the array-measured relative abundance of the *indica*- and *japonica-specific* transcript, which is denoted M.
4. A linear regression method is then applied on the input ratio (I) and the measured ratio M: $M = a + b \times I$, where a denotes intercept on the Y-axis and b means the slope.
5. To score the 12,950 (528 × 25) probe blocks on the rice tiling microarray, the correlation coefficient (r) between M and I is calculated for each block across the titration series.
6. A t test is then performed to examine whether an r value is significantly larger than its standard error, thus obtaining an FDR-adjusted P-value for each block.
7. Finally, the diagnostic blocks are selected by three criteria: (a) FDR P-value < 0.05; (b) $-1 < a < 1$; and (c) $b > 0.1$. Applying these criteria, at least one diagnostic block was identified for over half of the SNPs, albeit at different positions [29].

3.9.2 Determination of IAE in Reciprocal Rice Hybrids

The diagnostic probe blocks identified from the titration experiment can be used to measure allele-specific expression in a hybrid or a heterozygote. Further, the linear regression doubles as a standard curve to deduce quantitatively the relative levels of the allele-specific transcripts.

1. To identify IAE in the reciprocal hybrids between *japonica* and *indica*, the hybridization intensity is log2-transformed and quantile-normalized across the three replicates.
2. The measured ratio (M) is calculated in the reciprocal hybrids from each diagnostic probe blocks, and then the input ratio (I) is deduced based on M according to the two established parameters (a and b) in the regression equations.
3. For a given block, the log2-transformed I values from the three experimental replicates were used to perform a one-sample t test to examine if the mean differs from zero, with the null hypothesis that both alleles of a given gene contribute equally to the total amount of transcripts in the hybrid.
4. After obtaining a series of P values, they are ranked in ascending order: Rank (i) where $i = 1$ to n.
5. The P-values are then adjusted following the formula: FDR (i) = min(Pvalue(i) × n/Rank (i), FDR (i + 1)).
6. This process is reiterated as i is moved from n to 1 until FDR (n) = Pvalue (n).
7. Finally, the blocks with FDR-adjusted P-value less than 0.05 are selected as having detected IAE (*see* **Note 7**).

4 Notes

1. Because both the actual polymorphism and its sequence context will impact probe–target hybridization, a computation-based approach to select informative SNPs to quantitatively discriminate expressed alleles remains insufficient. One major goal of the tiling microarray analysis is to experimentally identify such SNPs.
2. High-quality RNA is required for optimal cDNA synthesis yield and cDNA labeling efficiency, which are critical to the success of microarray hybridization. Quantify each RNA sample according to the formula: RNA concentration (μg/ml) = A260 × 40 × dilution factor. A 260/280 ratio of 1.8–2.0 is a good indicator of RNA purity.
3. Ensure that the pH value of $NaHCO_3$ solution is no less than 9.0, which is crucial to successful sample labeling with the Cy3 dye.

4. From this step, care should be exercised to avoid direct exposure of samples to light (especially sunlight) because of photo-bleaching of fluorescence dye.
5. Proceed to the following operations, or store (up to 1 month) labeled samples at −15 to −25 °C protected from light.
6. Ensure that the total volume of probe mix is no less than 5 μl in case of diluting the hybridization buffer.
7. To validate microarray-detected IAE, a number of molecular assays can be performed. We recommend a simple allele-discriminating PCR method for this purpose. Three primers are designed for each pair of SNP-defined alleles, including two allele-discriminating forward primers and a common reverse primer [28]. Quantitative PCR should be performed using a SYBR Green kit. Following PCR, samples are subjected to melting curve analysis to confirm specificity of the amplicon. Relative abundance of allele-derived transcripts is obtained by normalizing against a control amplicon targeting the identical region between the two parental alleles.

References

1. Buckland PR (2004) Allele-specific gene expression in humans. Hum Mol Genet 13:255–260
2. Milani L, Gupta M, Andersen M et al (2007) Allelic imbalance in gene expression as a guide to *cis*-acting regulatory single nucleotide polymorphisms in cancer cells. Nucleic Acids Res 35:e34
3. Springer NM, Stupar RM (2007) Allele-specific expression patterns reveal biases and embryo-specific parent-of-origin effects in hybrid maize. Plant Cell 19:2391–2402
4. Khatib H (2007) Is it genomic imprinting or preferential expression? Bioessays 29:1022–1028
5. Yan H, Yuan W, Velculescu VE et al (2002) Allelic variation in human gene expression. Science 297:1143
6. Bray NJ, Buckland PR, Owen MJ et al (2003) *Cis*-acting variation in the expression of a high proportion of genes in human brain. Hum Genet 113:149–153
7. Lo HS, Wang Z, Hu Y et al (2003) Allelic variation in gene expression is common in the human genome. Genome Res 13:1855–1862
8. Ha G, Roth A, Bashashati A et al (2012) Integrative analysis of genome-wide loss of heterozygosity and mono-allelic expression at nucleotide resolution reveals disrupted pathways in triple negative breast cancer. Genome Res. doi:10.1101/gr.137570.112
9. Guo M, Rupe MA, Zinselmeier C et al (2004) Allelic variation of gene expression in maize hybrids. Plant Cell 16:1707–1716
10. Stupar RM, Springer NM (2006) *Cis*-transcriptional variation in maize inbred lines B73 and Mo17 leads to additive expression patterns in the F1 hybrid. Genetics 173: 2199–2210
11. Adams KL, Cronn R, Percifield R et al (2003) Genes duplicated by polyploidy show unequal contributions to the transcriptome and organ-specific reciprocal silencing. Proc Natl Acad Sci USA 100:4649–4654
12. Wang J, Tian L, Lee HS et al (2006) Genomewide nonadditive gene regulation in Arabidopsis allotetraploids. Genetics 172:507–517
13. Flagel L, Udall JA, Nettleton D et al (2008) Duplicate gene expression in allopolyploid Gossypium reveals two temporally distinct phases of expression evolution. BMC Biol 6:16
14. Jeong S, Hahn Y, Rong Q et al (2007) Accurate quantitation of allele-specific expression patterns by analysis of DNA melting. Genome Res 17:1093–1100
15. Winter E, Yamamoto F, Almoguera C et al (1985) A method to detect and characterize point mutations in transcribed genes: amplification and overexpression of the mutant c-Ki-ras allele in human tumor cells. Proc Natl Acad Sci USA 82:7575–7579
16. Singer-Sam J, Chapman V, LeBon JM et al (1992) Parental imprinting studied by allele-specific primer extension after PCR: paternal X chromosome-linked genes are transcribed prior to preferential paternal X chromosome

inactivation. Proc Natl Acad Sci USA 89:10469–10473
17. Skelly DA, Johansson M, Madeoy J et al (2011) A powerful and flexible statistical framework for testing hypotheses of allele-specific gene expression from RNA-seq data. Genome Res 21:1728–1737
18. Deveale B, van der Kooy D, Babak T (2012) Critical evaluation of imprinted gene expression by RNA-Seq: a new perspective. PLoS Genet 8:e1002600
19. Liljedahl U, Fredriksson M, Dahlgren A et al (2004) Detecting imbalanced expression of SNP alleles by minisequencing on microarrays. BMC Biotechnol 4:24
20. Ronald J, Akey JM, Whittle J et al (2005) Simultaneous genotyping, gene-expression measurement, and detection of allele-specific expression with oligonucleotide arrays. Genome Res 15:284–291
21. Udall JA, Swanson JM, Nettleton D et al (2006) A novel approach for characterizing expression levels of genes duplicated by polyploidy. Genetics 173:1823–1827
22. Yamada K, Lim J, Dale JM et al (2003) Empirical analysis of transcriptional activity in the Arabidopsis genome. Science 302: 842–846
23. Li L, Wang XF, Stolc V et al (2006) Genome-wide transcription analyses in rice using tiling microarrays. Nat Genet 38:124–129
24. Li L, He H, Zhang J et al (2008) Transcriptional analysis of highly syntenic regions between Medicago truncatula and Glycine max using tiling microarrays. Genome Biol 9:R57
25. Li L, Wang XF, Sasidharan R, Stolc V, Deng W, He H et al (2007) Global identification and characterization of transcriptionally active regions in the rice genome. PLoS One 2:e294
26. Li L, Wang XF, Xia M, Stolc V, Su N, Peng Z et al (2005) Tiling microarray analysis of rice chromosome 10 to identify the transcriptome and relate its expression to chromosomal architecture. Genome Biol 6:R52
27. Hazen SP, Naef F, Quisel T, Gendron JM, Chen H, Ecker JR et al (2009) Exploring the transcriptional landscape of plant circadian rhythms using genome tiling arrays. Genome Biol 10:R17
28. Kurihara Y, Kaminuma E, Matsui A, Kawashima M, Tanaka M, Morosawa T et al (2009) Transcriptome analyses revealed diverse expression changes in ago1 and hyl1 Arabidopsis mutants. Plant Cell Physiol 50:1715–1720
29. He H, Zhang H, Wang X, Wu N, Yang X, Chen R et al (2010) Development of a versatile, target-oriented tiling microarray assay for measuring allele-specific gene expression. Genomics 96:308–315
30. Yu J, Wang J, Lin W, Li S, Li H, Zhou J et al (2005) The genomes of *Oryza sativa*: a history of duplications. PLoS Biol 3:e38

Part III

Genetics and Epigenetics Regulation

Chapter 6

Detection of Epigenetic Alterations Using Tiling Arrays

Hoi-Hung Cheung, Owen M. Rennert, and Tin-Lap Lee

Abstract

The epigenetic status of cancer cells is a consequence of the neoplastic transformation of their normal counterpart. Epigenetic changes directly influence gene expression and chromatin organization, which consequently leads to escape from the tumor-suppression mechanisms. Global mapping for specific epigenetic modifications (e.g., DNA methylation) of the entire genome is required to reveal epigenetic hotspots associated with a cancer type/stage. DNA tiling arrays may be applied for genome-wide analysis of different epigenetic marks. Tiling arrays are high-density DNA microarrays that can be custom-made to survey regions of interest (e.g., gene promoters) or permit whole-genome analysis. To identify the genomic alterations associated with testicular cancers we used tiling arrays to profile their methylome. We successfully identified numerous epigenetically modified loci that arose as a consequence of tumor progression.

Key words Tiling array, MeDIP, Methylome, Epigenetic alteration

1 Introduction

Chemical modifications of the DNA backbone and histone residues of chromatin comprise the structural basis for complex control of genomic organization, stability, and transcriptional regulation. To identify genomic regions associated with a certain phenotype, for instance, transformation of normal cells to cancers, current techniques rely on the development of specific antibodies recognizing the unique modification in chromatin (ChIP). Thus, epigenetic alterations can be readily detected by comparison of the DNA abundance (enrichment) of the ChIP product using molecular techniques such as PCR or hybridization-based microarray (ChIP-chip) [1]. Next-generation sequencing (ChIP-seq) is another option for genome-wide studies.

DNA tiling arrays are high-density microarrays that can be customized, printed with millions of oligos on a single chip that permits genome-wide studies. Tiling arrays have been successfully applied for whole-genome profiling of histone modifications, transcriptome mapping, nucleosome positioning, copy number

Tin-Lap Lee and Alfred Chun Shui Luk (eds.), *Tiling Arrays: Methods and Protocols*, Methods in Molecular Biology, vol. 1067, DOI 10.1007/978-1-62703-607-8_6, © Springer Science+Business Media New York 2013

variation (CNV), and DNA methylation [2, 3]. We, and others, applied tiling array analysis for genomic profiling of DNA methylation in testicular tumors [4, 5]. First, immunoprecipitation was applied to enrich methylated DNA, using antibody specific for 5-methylcytidine, the only epigenetic modification on DNA. Second, the immunoprecipitated DNA (IP) was amplified, labeled, and hybridized to tiling arrays. Data analysis was based on the statistical difference in a defined window size between a set of probes from the IP and input samples. In this chapter, we describe the method for methylated DNA immunoprecipitation (MeDIP), and preparation of samples for hybridization to tiling arrays using the Affymetrix platform (Fig. 1). This protocol has been successfully demonstrated to identify regions of differentially methylated regions (DMR) in tumors [4].

2 Materials

2.1 MeDIP

1. Antibody: Mouse monoclonal anti-5-Methylcytidine, Clone 33D3 (Eurogentec).
2. Sheep anti-mouse IgG conjugated Dynabeads (Life Technologies).
3. Sonicator: Ultrasonic Processor Model GE505 (Sonics).
4. 10× IP buffer: 100 mM sodium phosphate (pH 7.0), 1.4 M sodium chloride, 0.5 % Triton X-100. Dilute tenfold with nuclease-free water to give 1× IP buffer.
5. PBS–BSA 0.1 %: Dilute BSA in 1× PBS to 0.1 % (w/v).
6. Magnetic rack for Dynabeads (Life Technologies).
7. Proteinase K (Roche).
8. Proteinase K digestion buffer: 50 mM Tris–HCl (pH 8.0), 10 mM EDTA, 0.5 % SDS.
9. Phenol:chloroform:isoamyl alcohol (25:24:1, v/v).
10. Glycogen (Life Technologies).

2.2 Preparation of DNA Probes

1. Primer A: GTTTCCCAGTCACGGTC$(N)_9$, HPLC purified, diluted to 200 μM in TE buffer.
2. Primer B: GTTTCCCAGTCACGGTC, HPLC purified, diluted to 100 μM in TE buffer.
3. Sequenase Version 2.0 DNA Polymerase and 5× Sequenase Reaction Buffer (USB).
4. Platinum Taq and 10× PCR Buffer (Life Technologies).
5. dNTP/dUTP mix: 10 mM dATP, 10 mM dCTP, 10 mM dGTP, 8 mM dTTP, 2 mM dUTP (Roche).
6. PCR Thermocycler (Applied Biosystems).

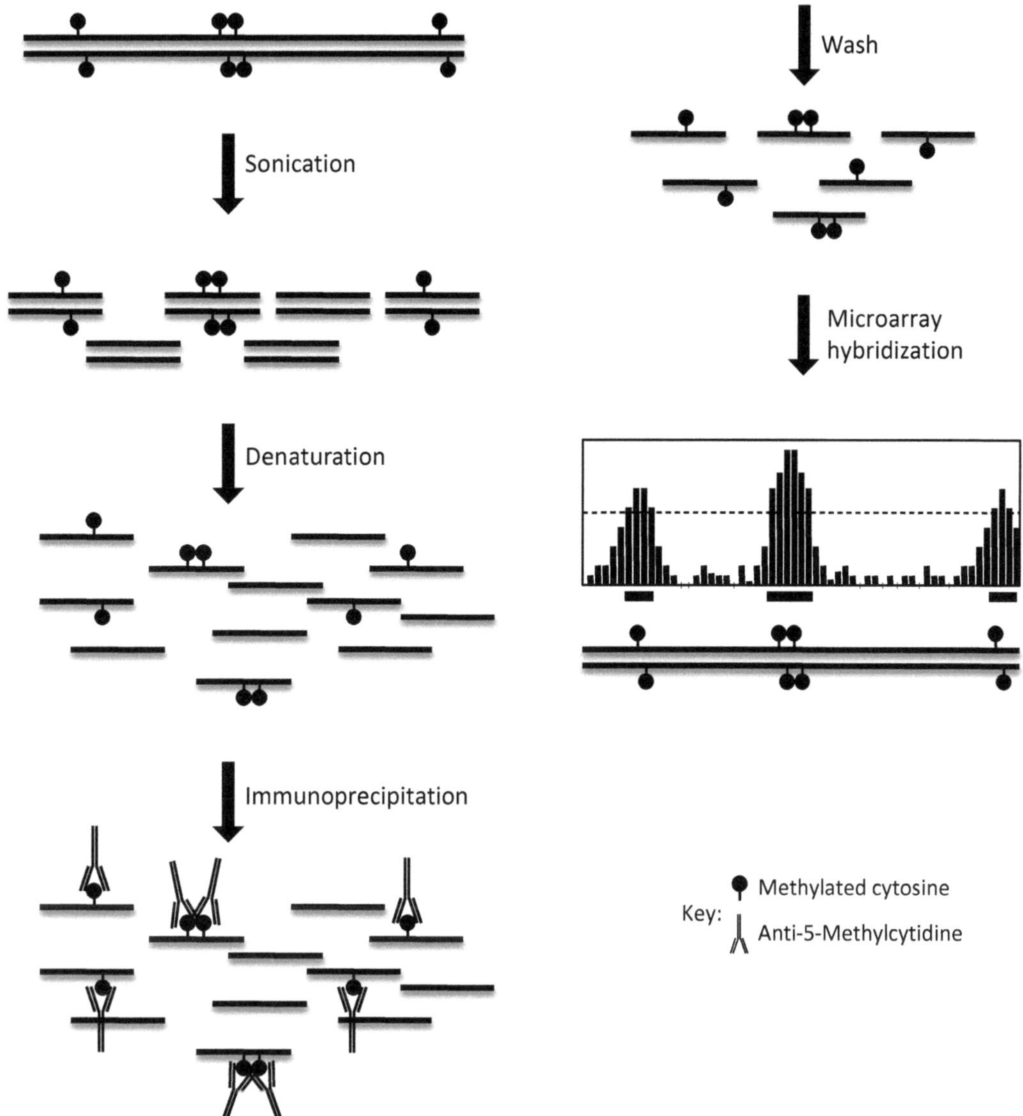

Fig. 1 Overview of MeDIP-chip. Genomic DNA is sonicated to random fragments. DNA fragments are heat-denatured, and then captured by anti-5-methylcytidine antibodies. After washing, methylated genomic fragments are enriched, amplified, and hybridized to tiling array. Methylated loci are represented by *peaks* above the cutoff as shown in Integrated Genomic Browser

7. PrepEase DNA Clean-Up Kit (USB) or MinElute Reaction Cleanup Kit (Qiagen).
8. Uracil-DNA Glycosylase (UDG), 2 U/μl (USB).
9. Human apurinic/apyrimidinic endonuclease 1 (APE 1), 10 U/μl, and 10× APE 1 reaction buffer (USB).
10. Terminal deoxynucleotidyl transferase (rTdT), 30 U/μl, and 5× TdT buffer (USB).
11. DNA-labeling reagent, 10 mM (USB).

2.3 Hybridization to Tiling Arrays

1. Human Tiling 2.0R Array Set or Human Promoter 1.0R Array (Affymetrix).
2. GeneChip Hybridization, Wash, and Stain Kit (Affymetrix).
3. Control Oligo-B2, 3 nM (Affymetrix).

2.4 Instruments for Tiling Array Hybridization

1. GeneChip Hybridization Oven 640 (Affymetrix).
2. GeneChip Fluidics Station 450 or 400 (Affymetrix).
3. GeneChip Scanner 3000 7G (Affymetrix).

3 Methods

3.1 MeDIP

1. Prepare high-grade genomic DNA samples by appropriate methods of DNA extraction (*see* **Note 1**).
2. Adjust ~5–6 μg of DNA in TE buffer to a final volume of 400 μl.
3. Chill DNA on ice for ~5–10 min.
4. Perform fragmentation of DNA by sonication on ice or other chilled condition (*see* **Note 2**).
5. Check the size of fragmented DNA on a 1.5 % agarose gel by electrophoresis.
6. Determine the concentration of the fragmented DNA.
7. Adjust 5 μg of the fragmented DNA in TE buffer to a final volume of 450 μl in a nuclease-free microcentrifuge tube.
8. Denature the DNA at 95 °C for 10 min in a heat-block.
9. Cool the DNA immediately on ice for ~2 min.
10. Add 51 μl of 10× IP buffer and 10 μl of 5mC antibody.
11. Mix thoroughly. Incubate at 4 °C for 2 h using an overhead shaker.
12. Prepare Dynabeads (30 μl of each IP) in microcentrifuge tubes.
13. Remove the storage buffer after magnetic precipitation.
14. Wash the beads twice with 800 μl of PBS–BSA 0.1 %.
15. Resuspend the beads in 30 μl of 1× IP buffer.
16. Add the beads to the IP mix (**step 11**). Incubate at 4 °C for 2 h using an overhead shaker.
17. Wash the beads three times with 700 μl of IP buffer (*see* **Note 3**).
18. Remove IP buffer and resuspend the beads in 250 μl digestion buffer.
19. Add 5 μl of Proteinase K (20 mg/ml). Seal the tube with parafilm.
20. Digest on a heat-block with orbital shaking (800 rpm, 50 °C for 3 h).

21. After digestion, purify DNA using phenol/chloroform/isoamyl alcohol method or other commercially available DNA purification columns (*see* **Note 4**).
22. To precipitate DNA, add 500 µl of absolute ethanol, 20 µl of 5 M NaCl, and 1 µl of glycogen (20 µg/µl).
23. Mix and incubate at −20 °C for 4 h or overnight.
24. Centrifuge at 20,000 × *g* for 30 min.
25. Discard the supernatant and wash the DNA pellet with 700 µl of 70 % ethanol. Centrifuge at 20,000 × *g* for 10 min.
26. Discard the ethanol and air-dry the pellet at room temperature for ~10 min.
27. Resuspend the dried DNA in 10 µl of nuclease-free water.
28. (Optional) Dilute 2 µl of the immunoprecipitated DNA in 18 µl of TE buffer. Check the efficiency of MeDIP by qPCR using primers for known methylated (positive controls) and unmethylated loci (negative controls).

3.2 Preparation of DNA Probes

3.2.1 PCR Amplification

1. In PCR tubes, set up the random priming reaction and mix thoroughly:

MeDIP DNA	8 µl
5× Sequenase reaction buffer	4 µl
Primer A (200 µM)	4 µl
Nuclease-free water	1.15 µl
Total	17.15 µl

2. In a PCR thermocycler, heat the samples at 95 °C for 4 min. Hold the program and snap cool the samples on ice for 2 min.
3. Add the following cocktail to the PCR tube and mix thoroughly:

DTT (0.1 M)	1 µl
BSA (10 mg/ml)	0.2 µl
Sequenase (1.3 U/µl)	1 µl
dNTP mix (10 mM each)	1.25 µl
Total	3.45 µl

4. In a PCR thermocycler, incubate the samples at 10 °C for 5 min. Then ramp the samples slowly from 10 to 37 °C for 10 min with the lip of the machine opened. Hold at 37 °C for 8 min.
5. Repeat three times for **steps 2–4** with 1 µl of fresh Sequenase added in **step 3**.

6. Fill the reaction mix to 50 µl by adding 26.4 µl of TE buffer.
7. Set up the PCR mix:

DNA template (**step 6**)	30 µl
10× PCR buffer	20 µl
dNTP/dUTP mix	7.5 µl
Primer B (100 µM)	8 µl
DMSO	12 µl
$MgCl_2$ (50 mM)	8 µl
Platinum Taq (5 U/µl)	4 µl
Nuclease-free water	110.5 µl
Total	200 µl

Run PCR program: Initial denaturation 92 °C/2 min; 92 °C/30 s, 40 °C/30 s, 50 °C/30 s, 72 °C/1 min (30 cycles); final extension 72 °C/5 min.

8. Purify the PCR products with commercially available DNA cleanup columns (*see* **Note 5**).

3.2.2 Fragmentation and Labeling of Probes

1. In PCR tubes, set up fragmentation reaction mix for each MeDIP/input sample:

Purified DNA	9 µg
10× APE 1 reaction buffer	4.8 µl
UDG (2 U/µl)	4 µl
APE 1 (10 U/µl)	7 µl
Nuclease-free water filled up to 48 µl	
Total	48 µl

2. Mix well and incubate at 37 °C for 1 h, 93 °C for 2 min, and 4 °C for ≥2 min (*see* **Note 6**).
3. In PCR tubes, set up labeling reaction mix:

Fragmented DNA	45 µl
5× TdTbuffer	12 µl
rTdT (30 U/ul)	2 µl
DNA-labeling reagents (10 mM)	1 µl
Total	60 µl

4. Mix well and incubate at 37 °C 1 h, 70 °C for 10 min, and 4 °C for ≥2 min (*see* **Note 7**).

3.3 Hybridization to Tiling Arrays

1. In microcentrifuge tubes, prepare the hybridization mix for each MeDIP/input sample:

Labeled DNA	60 μl
Control oligo-B2	4 μl
2× Hybridization mix	120 μl
DMSO	16.8 μl
Nuclease-free water	39.3 μl
Total	240 μl

2. Mix well and incubate at 99 °C for 5 min. Then cool to 45 °C for 5 min.
3. Centrifuge at maximum speed for 1 min to remove any precipitate.
4. Prehybridize tiling array chips with 200 μl of prehybridization solution at 45 °C for 5 min.
5. Remove prehybridization solution from the chips and replace with 200 μl of hybridization mix. Seal the septa with sticky labels to prevent leakage.
6. Perform hybridization in a hybridization oven set to 45 °C, 60 rpm for 16 h.
7. After overnight hybridization, remove and save the hybridization mix. Proceed to the array washing, staining, and scanning procedures according to the manufacturer's instructions (*see* **Note 8**).
8. Raw data in CEL format from tiling arrays can be processed by TAS (Affymetrix) or other publicly available software such as CisGenome [6]. The downstream analysis can be further visualized in genome browsers such as IGB and UCSC Genome Browser.

4 Notes

1. It is recommended that at least 4–5 μg of DNA is available for MeDIP. Genomic DNA should not be modified, amplified, or degraded before MeDIP.
2. Fragment size ranging from 100 to 500 bp is optimal for MeDIP. For instance, when using the listed model of sonicator, we suggest the following condition: pulser with 15 s ON and 15 s OFF at 20 % of amplitude for 10 min. Overheating or prolonged sonication results in loss of DNA.
3. To wash the IP mix, first precipitate the beads with a magnetic rack for 2 min. Carefully remove the aqueous supernatant and

then suspend in 700 μl of IP buffer. Wash the beads in an overhead shaker for 5 min at room temperature. Repeat the washing steps for another two times.

4. To purify DNA, add 500 μl of phenol/chloroform/isoamyl alcohol and mix vigorously by vortexing for 30 s. Centrifuge at 12,000 × *g* for 2 min and then transfer the supernatant to a fresh tube. Repeat once for phenol/chloroform/isoamyl alcohol extraction.
5. Cleanup of PCR products can be achieved using PrepEase DNA Clean-Up Kit (USB) or MinElute Reaction CleanupKit (Qiagen) according to the manufacturer's instructions.
6. To check the efficiency of fragmentation, take 1–2 μl of the reaction mixture and run the samples in a Bioanalyzer.
7. To check the efficiency of labeling, take 1–2 μl of the reaction mixture and run the samples using gel-shift assay.
8. A complete procedure for array washing, staining, and scanning can be found from the manufacturer's user manual (http://media.affymetrix.com/support/downloads/manuals/wash_stain_scan_cartridge_arrays_manual.pdf.

References

1. Tollefsbol TO (2004) Methods of epigenetic analysis. In: Tollefsbol TO (ed) Methods Mol Biol 287:1–8, Humana, Totowa, NJ
2. Liu XS (2007) Getting started in tiling microarray analysis. PLoS Comput Biol 3:1842–1844
3. Zilberman D, Henikoff S (2007) Genome-wide analysis of DNA methylation patterns. Development 134:3959–3965
4. Cheung HH, Lee TL, Davis AJ, Taft DH, Rennert OM, Chan WY (2010) Genome-wide DNA methylation profiling reveals novel epigenetically regulated genes and non-coding RNAs in human testicular cancer. Br J Cancer 102:419–427
5. Mockler TC, Chan S, Sundaresan A, Chen H, Jacobsen SE, Ecker JR (2005) Applications of DNA tiling arrays for whole-genome analysis. Genomics 85:1–15
6. Ji H, Jiang H, Ma W, Johnson DS, Myers RM, Wong WH (2008) An integrated software system for analyzing ChIP-chip and ChIP-seq data. Nat Biotechnol 26:1293–1300

Chapter 7

Investigating Gene Promoter Methylation in a Mouse Model of *Status Epilepticus*

Suzanne F.C. Miller-Delaney, Sudipto Das, Raymond L. Stallings, and David C. Henshall

Abstract

Epigenetic modification of DNA by methylation of the cytosine present in CG dinucleotides constitutes a key regulatory mechanism in the control of gene expression in neurological diseases. In this chapter, we describe an in-depth methodology of methylated DNA immunoprecipitation used in combination with tiling microarrays (MeDIP-chip) in order to analyze genome-wide gene promoter methylation in the hippocampus of mice following *status epilepticus* (prolonged seizure). While a specific mouse model and array format are described, the method can be applied to DNA from many tissues to analyze the methylation status of promoter regions across whole genomes, using a wide range of available array formats (both custom designed and commercially catalogued). We conclude the chapter with the description of bisulfite sequencing validation of MeDIP-chip results.

Key words Epigenetics, DNA methylation, Promoter methylation, Methylated DNA immunoprecipitation (MeDIP), *Status epilepticus*

1 Introduction

DNA methylation is defined as the covalent attachment of methyl groups to the cytosine present in CG dinucleotide-containing regulatory sequences, a process catalyzed by DNA methyl transferases (DNMTs) [1, 2]. While originally thought to be a static process after cellular differentiation, methylation of gene promoter regions is responsible, in part, for the modulation of gene expression profiles that promote cell identity and function throughout life [3]. In the brain, DNA methylation has been shown to be highly dynamic in the hippocampus in response to neural activity and is important for certain plasticity and injury responses [4–9]. In fact, aberrant methylation profiles have been associated with differential gene expression in neurodevelopmental, neuropsychiatric, and neurodegenerative disorders [4, 7] including temporal lobe epilepsy, where

Tin-Lap Lee and Alfred Chun Shui Luk (eds.), *Tiling Arrays: Methods and Protocols*, Methods in Molecular Biology, vol. 1067, DOI 10.1007/978-1-62703-607-8_7, © Springer Science+Business Media New York 2013

DNMT1 and 3a levels are increased [10] and more specifically, differential hypermethylation of the *reelin* promoter has been directly correlated with reduced transcription of the gene [11].

In this chapter, we detail the use of methylated DNA immunoprecipitation (MeDIP) in combination with tiled microarrays covering all annotated gene promoters in the genome (MeDIP-chip) to analyze differential gene promoter methylation in a mouse model of *status epilepticus* [6]. While *status epilepticus* is a prolonged seizure state, it is nonetheless useful for studying how seizures can damage the brain and lead to epilepsy. Indeed, *status epilepticus* can be profoundly damaging to the brain, producing pathophysiological changes resulting in widespread neuronal loss and reactive gliosis which has been associated with large-scale, wide-ranging changes in gene expression [12–16]. Here we employ a mouse model of *status epilepticus* where focally evoked limbic seizures are induced by unilateral intra-amygdala injection of 1 μg kainic acid (KA), an agonist for the kainate subtype of glutamate receptors which are found at their highest density in the amygdala and the CA3 region of the hippocampus [6, 14–17]. *Status epilepticus* is terminated in the model by intra-peritoneal administration of Lorazepam. Seizures result in a near-complete CA3 lesion, with minor injury to the CA1 and hilus [6, 14–16]. Mice go on to develop spontaneous recurrent seizures (i.e., epilepsy) within 3–5 days and hallmark features of temporal lobe epilepsy [18].

MeDIP-chip involves the selective immunoprecipitation of methylated cytosines using an antibody to 5-methylcytosine, followed by differential labeling of MeDIP and Input DNA, and subsequent hybridization to microarrays (*see* Fig. 1). The first high-resolution map of the human methylome was generated using this protocol in 2005 [19]. In addition to use on microarrays (as described here), MeDIP can be used in combination with several platforms including whole-genome sequencing (MeDIP-seq). MeDIP-chip experimental results are routinely validated using bisulfite sequencing, a relatively simple and cost effective technique [6, 20–22]. Bisulfite sequencing takes advantage of the differing sensitivity of cytosine and methyl-cytosine to sodium bisulfite under acidic conditions, in which cytosine (C) undergoes conversion to uracil (T), whereas methylated cytosine remains unreactive. Following bisulfite conversion of DNA, a genomic region of interest can be PCR amplified and the product sequenced in order to analyze the methylation status of the region, therefore validating MeDIP-microarray results on a gene to gene basis [6, 20–22]. We conclude the chapter with a detailed description of the design and implementation of bisulfite sequencing, focusing on a single gene, ATPase Ca^{2+}-sequestering (*Atp2c1*) for illustrative purposes.

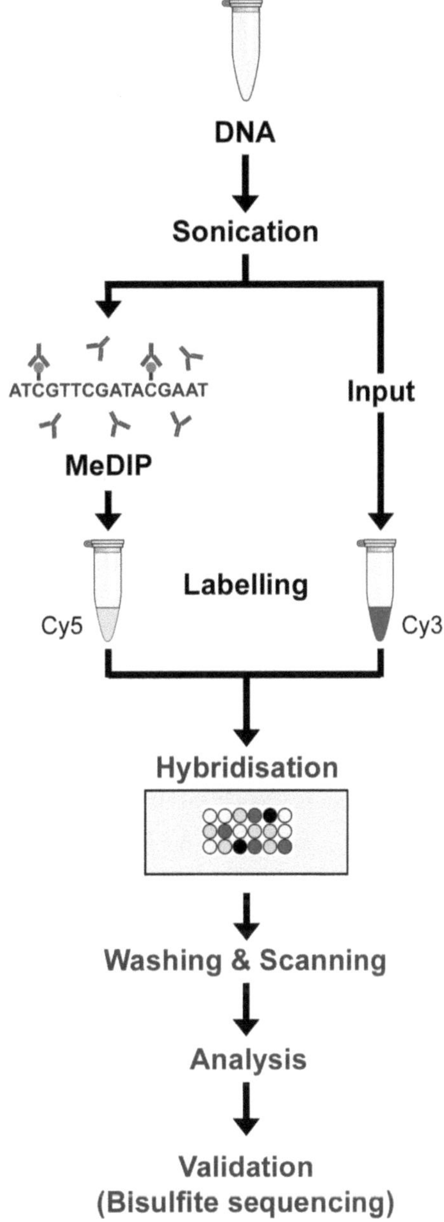

Fig. 1 Flowchart illustrating MeDIP-chip experimental paradigm

2 Materials

2.1 Preparation of Animal Model and DNA Extraction from Hippocampus

1. Ethical approval from the competent authority is required for animal experiments, which can only be undertaken under license from the appropriate authority. In our case, animal experiments are carried out in accordance with the European Communities Council Directive (86/609/EEC) and are

reviewed and approved by the Research Ethics Committee of the Royal College of Surgeons in Ireland, under license from the Department of Health, Dublin, Ireland.

2. C57BL/6 mice (Harlan) 20–25 g, housed in a climate-controlled vivarium on a 12 h light/dark cycle with food and water provided ad libitum.
3. Intraperitoneal injection; 26 gauge needles, 1 ml syringe.
4. Intra-amygdala injection; isoflurane (3–5 %) with appropriate scavenging system, feedback-controlled heat blanket (Harvard Apparatus), feedback-controlled incubator, observation chamber, stereotaxic frame suitable for mouse surgery, guide cannulas (26 G) and injection cannulas (31G) (both from Plastics One), dental cement.
5. Kainic Acid (KA; Sigma); 1 μg in 0.2 μl for intra-amygdala micro-injection.
6. Lorazepam; 6 mg/kg in PBS for intraperitoneal injection.
7. 1.5 ml tubes.

2.2 Preparation of DNA

1. DNeasy Blood and Tissue kit (Qiagen) or similar.
2. Nanodrop spectrophotometer.
3. Vacuum centrifuge.
4. Nuclease-free molecular grade water (dH_2O).
5. RNase A (100 mg/ml).
6. Bench top centrifuge.
7. Sonicator suitable for use with 1.5 ml tubes.
8. 2 % Agarose Gel and electrophoresis system (E-gel, Invitrogen) or similar.
9. 123bp DNA ladder (Sigma) or similar.
10. An appropriate gel documentation system.

2.3 Methylated DNA Immunoprecipitation (MeDIP)

1. TE buffer; 10 mM Tris–HCl, pH 7.5, 1 mM EDTA.
2. Round-bottomed 2 ml tubes.
3. Orbital rotator at 4 °C.
4. Monoclonal antibody against 5-methylcytidine (Eurogentec).
5. 10× IP Buffer; 100 mM Sodium Phosphate pH 7.0, 1.4 M Sodium Chloride, 0.5 % Triton X-100.
6. Shaking bench-top heat block.
7. Dynabeads (Sheep anti-mouse IgG) and magnetic particle concentrator (Invitrogen).
8. BSA; 0.1 % in PBS.
9. Proteinase K Buffer; 50 mM Tris–HCl 8.0, 10 mM EDTA, 0.5 % SDS (this must be made up fresh each time).

10. Proteinase K (10 mg/ml stock).
11. Phenol chloroform extraction (phenol, chloroform, 5 M NaCl, glycogen, 100 % ethanol, 80 % ice-cold ethanol, molecular grade dH_2O).
12. Vortex, bench-top centrifuge capable of 13,000 × *g* and 1.5 ml tubes.

2.4 Fluorescent Labelling of Samples

1. Nanodrop spectrophotometer or similar.
2. NimbleGen Dual Color DNA labeling kit (Roche Nimblegen).
3. Molecular grade dH_2O.
4. Isopropanol, 80 % ice-cold ethanol.
5. Vortex, bench-top centrifuge capable of 13,000 × *g* and 1.5 ml tubes.
6. Vacuum centrifuge.

2.5 Hybridization, Washing and Scanning of Arrays

1. Nanodrop spectrophotometer or similar.
2. Methylation-specific tiling arrays covering all annotated promoter regions in the mouse genome. In this case we use Mouse Meth 385 K Prom Plus CpG arrays (Roche Nimblegen) with appropriate mixers and Precision Mixer Alignment tool (PMAT) (*see* **Note 1**).
3. Hybridization kit and Wash Buffer Kit (Roche Nimblegen).
4. Vortex and bench-top centrifuge.
5. MAUI hybridization system (BioMicro Systems).
6. ArrayIt microarray high-speed centrifuge or other slide dryer.
7. Axon 4000B scanner and GenePix Pro 6.0 software.
8. Software for analysis and visualization of array data. In this case, Nimblescan version 2.4 and SignalMap 1.9 (Roche Nimblegen).

2.6 Bisulfite Sequencing

1. Nanodrop spectrophotometer and vacuum centrifuge.
2. 500 ng DNA from the CA3 region of three individual control and SE mice.
3. Bisulfite conversion; EZ DNA methylation Gold kit (Zymo Research), thermocycler.
4. PCR; PCR tubes, Pyromark PCR kit (Qiagen), custom primers for gene of interest (In the example shown, primers used are forward primer, 5′-GTATTTGTAAGAGAAATTAGGAGAAG-3′ and reverse primer, 5′-TATCTACTCCTACCCCTATTTCC-3′).
5. Visualisation of PCR products; 2 % Agarose Gel and electrophoresis system (E-gel, Invitrogen) or similar, DNA ladder (Sigma), gel documentation system.

6. Purification of PCR products; QIAquick PCR purification kit (Qiagen).
7. Sequencing facilities. In this case we use Eurofins MWG Operon.

3 Methods

3.1 Preparation of the Mouse Model

Animal experiments are performed as described previously [6, 14, 15]. *Status epilepticus* is induced by intra-amygdala microinjection of KA. Non-seizure controls undergo intra-amygdala injection of PBS.

1. Anesthetize mice using isoflurane (3–5 %) and maintain normothermic by means of a feedback-controlled heat blanket.
2. Place mice in a stereotaxic frame and following a midline scalp incision perform a full craniectomy.
3. A guide cannula is affixed using dental cement (coordinates from Bregma: AP = −0.94; L = −2.85 mm) [23].
4. Discontinue anesthesia and remove mice from stereotaxic frame. Place in an observation chamber until the mice recover.
5. Lightly restrain mice and lower an injection cannula 3.75 mm below the dura for injection of KA or PBS into the basolateral amygdala nucleus.
6. Place animals in a feedback-controlled incubator at 37 °C to recover.
7. Administer Lorazepam (6 mg/kg in 0.2 ml, intraperitoneal) to all animals 40 min after intra-amygdala KA to minimize morbidity and mortality from *status epilepticus.*
8. Sacrifice animals 24 h following intra-amygdala KA injection by cervical dislocation.
9. Microdissect the CA3 subfield of the hippocampus [24] on ice/cool block and place tissue in a sterile 1.5 ml tube. Flash freeze in liquid nitrogen for temporary storage at −80 °C (*see* **Note 2**).

3.2 Preparation of DNA

1. DNA is isolated from hippocampal CA3 subfields using the DNeasy Blood and Tissue kit (Qiagen). Simultaneous extraction of protein, RNA and DNA can be carried out if required for downstream application (*see* **Note 3**).
2. Calculate DNA concentration using a nanodrop spectrophotometer. Depending on the extraction method used, up to six CA3 subfields need to be pooled in order to obtain 5 μg DNA. Duplicate arrays with independent biological replicates should be set up for each condition.

3. Prepare 5 μg DNA with nuclease-free water to a final volume of 100 μl (vacuum centrifuge at 30 °C to reduce volume if necessary).
4. Treat DNA with RNAse A (2 μl per tube) and let sit at room temperature for 4 min (*see* **Note 4**).
5. Vortex the solution and centrifuge briefly.
6. Sonicate each sample seven times, for 10 s each placing the sample on ice for 1 min between each sonication (*see* **Note 5**). All sonication steps are carried out at 50 % amplitude and at 1 cycle.
7. Resolve 100 ng of sonicated DNA on a 2 % agarose gel (in this case the E-gel system) with an appropriate DNA ladder.
8. Visualize the DNA on a gel documentation system. If sonicated sufficiently, the DNA should be resolved to between 300 and 800 bp (*see* Fig. 1). If required, repeat 1–2 cycles of sonication and check on gel until DNA is sonicated sufficiently.

3.3 Methylated DNA Immunoprecipitation (MeDIP)

1. 1.5–4 μg DNA is diluted in TE buffer to a total volume of 450 μl and placed in a round-bottomed 2 ml tube (A portion of sonicated DNA should be left untreated to serve as an input control). The use of round-bottomed tubes is critical as it prevents settling of Dynabeads during immunoprecipitation.
2. Following denaturation for 10 min at 98 °C in a heat block, 51 μl of 10× IP buffer is added (*see* **Note 6**).
3. Add 10 μl of 5-methylcytidine antibody to the denatured DNA (*see* **Note 7**) and incubate overnight at 4 °C on a tube rotator (16 h).
4. Pre-wash 40 μl of Dynabeads (per sample) with 800 μl PBS–BSA 0.1 % for 5 min at room temperature with vigorous shaking.
5. Collect the beads with a magnetic particle concentrator and repeat the wash step with additional 800 μl PBS–BSA 0.1 %.
6. Collect the beads with a magnetic particle concentrator and resuspend in 40 μl of 1× IP buffer.
7. Add dynabeads to the sample and incubate for 3 h at 4 °C on a tube rotator.
8. Collect the beads with a magnetic particle concentrator and wash with 700 μl 1× IP buffer for 10 min at room temperature with vigorous shaking.
9. Repeat the wash twice with 700 μl 1× IP buffer, pipetting up and down ten times in order to ensure thorough washing of beads. Care should be taken that no air-bubbles are introduced during mixing.

10. Collect the beads with a magnetic particle concentrator and resuspend in 250 μl proteinase K digestion buffer.
11. Add 7 μl of proteinase K and incubate for 2–3 h at 50 °C with shaking.
12. Add one volume of Phenol (250 μl), vortex and centrifuge at 16,000 × g for 5 min.
13. Remove the top phase into a fresh 1.5 ml tube and add one volume of chloroform (250 μl). Vortex and centrifuge at 16,000 × g for 5 min.
14. Precipitate the DNA with 20 μl 5 M NaCl, 1 μl glycogen and 500 μl 100 % ethanol. Vortex following the addition of each component and centrifuge at 13,000 × g for 10 min.
15. Remove the supernatant and resuspend pellet in 200 μl ice-cold 80 % ethanol, Vortex and centrifuge at 13,000 × g for 10 min.
16. Remove the supernatant and vacuum centrifuge at low heat (30 °C) for 5 min to dry the pellet.
17. Resuspend the DNA pellet in 60 μl dH_2O and store at −20 °C.

3.4 Flourescent Labelling of DNA Samples

1. Briefly centrifuge and resuspend fluorescent primers in the random primer buffer (with added β-mercaptoethanol) provided with the NimbleGen Dual Color DNA labeling kit. Aliquot in volumes of 41 μl and store at −20 °C (use within 4 months).
2. Pipette 10 μl of input sample (20 ng/μl; 200 ng) into a 1.5 ml tube and add 30 μl dH_2O.
3. Measure the concentration of MeDIP samples using a nanodrop spectrophotometer.
4. Pipette the appropriate volume of MeDIP sample into a 1.5 ml tube (200 ng). Vacuum centrifuge or add dH_2O to a maximum volume of 40 μl if necessary.
5. Add 40 μl fluorescent primers to samples. In this example we add Cy3 to input sample and Cy5 to MeDIP sample.
6. Heat-denature samples for 10 min at 98 °C and then place on ice for at least 2 min.
7. Prepare the Klenow mastermix (included in the NimbleGen Dual Color DNA labeling kit) and add 20 μl to each sample on ice.
8. Mix well by pipetting up and down ten times (do not vortex).
9. Centrifuge briefly and incubate overnight at 37 °C, protected from light.
10. Stop the reaction by the addition of 21.5 μl EDTA (provided with the NimbleGen Dual Color DNA labeling kit).

11. Vortex briefly and transfer to tube containing 110 μl isopropanol.
12. Vortex, centrifuge briefly and incubate for 10 min at room temperature, protected from the light.
13. Centrifuge at 12,000 × *g* for 10 min. Carefully tip off the supernatant and rinse the pellet by adding 500 μl ice-cold 80 % ethanol.
14. Centrifuge at 12,000 × *g* for 2 min and carefully tip of the supernatant.
15. Dry pellet in a vacuum centrifuge and rehydrate in 25 μl dH_2O.
16. Store at −20 °C, protected from light for up to 1 month (*see* **Note 8**).

3.5 Hybridization of Samples to Microarrays

1. The concentration of labeled MeDIP and input sample is measured using a nanodrop spectrophotometer.
2. Turn on the MAUI hybridization system and set to 42 °C (with the cover closed, it can take up to 3 h for the temperature to stabilize).
3. Following consultation with protocols according to array format, combine the relevant concentration of labeled MeDIP and input sample. In the case of one-plex 385 K arrays, 6 μg of MeDIP and input sample are combined and vacuum centrifuged at 30 °C to a maximum volume of 5 μl. Where necessary, molecular grade dH_2O can be used to bring the volume of sample up to 5 μl.
4. Prepare the hybridization mix (Nimblegen) to a total volume of 17 μl (11.8 μl hybridization buffer; 4.7 μl component A and 0.5 μl alignment oligo). Vortex well for 15 s and spin using on a bench-top centrifuge.
5. Denature samples at 95 °C for 5 min (protected from light) and place tubes at 42 °C for at least 5 min or until ready.
6. Prepare and affix the appropriate mixer to the microarray. In this case, we use Nimblegen-specific single-plex formats alongside the PMAT tool. The mixer-microarray assembly is then placed in the slide bay of the preheated Maui Hybridization system.
7. Load the appropriate volume of sample (18 μl in this case) into the fill port of the mixer, referring to manufacturers instructions throughout (*see* **Note 9**).
8. Dry any overflow from and seal the fill and vent ports using lint free tissue.
9. Close the hybridization bay and turn on the mixing panel on the Maui system, ensuring the mix mode is set to B. Press the

mix button to start mixing. The Maui system should recognize the slides in occupied bays (as shown by a green indicator light). Double check the Maui system 10 min later to ensure the system in running correctly.

10. Hybridize samples at 42 °C overnight (16–20 h).

3.6 Washing and Scanning of Microarrays

1. Prepare washes according to manufacturers' instructions (in this case we use the Nimblegen Wash Buffer kit). The first wash should be prepared to 42 °C and be of sufficient volume to submerge the microarray completely in order to remove the adhered mixer. Microarrays should be processed one at a time and brought through each washing solution without interruption (*see* **Note 10**). When processing each microarray, check the temperature of wash I in order to avoid array damage by forced mechanical removal of the mixer.
2. Spin-dry washed microarrays for 1 min on the ArrayIt microarray high-speed centrifuge and blot edges with tissue in order to remove any residual liquid. Store the dried slide in a dark dessicator until ready to scan.
3. Scanning of arrays is performed on an Axon 4000B scanner and data processed using GenePix Pro 6.0. Image analysis.
4. Detection of hypermethylation peaks is performed using the methylation application in NimbleScan version 2.4. Methylated peaks are identified using the following parameters: sliding window of 750 bp, p value minimum cutoff (−log10) of 2.0, and a minimum of two probes per peak. Only hypermethylated peaks that are detected in replicate arrays are subsequently carried forward for analysis.

3.7 Verification of Array Results by Bisulfite Sequencing

1. Additional animal samples may be prepared (if required) for the verification of methylation status and DNA extracted from the CA3 region of the hippocampus as previously described (*see* Subheading 3.2). Ideally, sufficient amounts of DNA remain from the original tissue for use in subsequent validation of array results.
2. Calculate the concentration of DNA using a nanodrop spectrophotometer and place 500 ng in a sterile 1.5 ml tube. The maximum volume of DNA samples should not exceed 20 μl (Vacuum centrifuge if necessary to reduce volume).
3. Add 130 μl of conversion reagent (EZ DNA methylation Gold kit, Zymo Research) per sample. Divide the solution into two 75 μl PCR tubes and convert over 5 h in thermocycler (98 °C for 10 min, 53 °C for 5 h).

4. Load both tubes of sample into a single Zymo-spin IC column (EZ DNA methylation Gold kit, Zymo Research) and wash using the accompanying buffers.
5. Elute bisulfite converted DNA in 10 µl molecular grade dH_2O. Store at −80 °C (*see* **Note 11**).
6. Design PCR primers to genes of interest using methyl primer express (www.appliedbiosystems.com/methylprimerexpress) (*see* Fig. 2). Primers should amplify a product of 100–400 bp and should be positioned around the methylation peaks of interest. CG dinucleotides should be avoided in the primer design. In the example shown, primers for Atp2c1 are as follows; forward primer, 5′-GTATTTGTAAGAGAAA TTAGGAGAAG-3′ and reverse primer, 5′-TATCTACTCC TACCCCTATTTCC-3′ (*see* Fig. 2) and amplify a product of 240 bp.
7. Calculate the concentration of bisulfite converted DNA using a nanodrop spectrophotometer (on a single-stranded DNA setting) and carry out PCR on 10 ng of bisulfite-treated DNA using the Pyromark PCR kit and self-designed primers. PCR should be carried out in a mastermix volume of 50 µl and adjust annealing temperature according to primer design (in this case of Atp2c1, an annealing temperature of 54.8 °C was used). Where possible always include a positive and negative control in your pcr design.
8. Resolve 10 µl of pcr product on a 2 % agarose gel (in this case the E-gel system) with an appropriate DNA ladder. Visualize the DNA on a gel documentation system.
9. Purify the remaining PCR products using the QIAquick PCR purification kit (Qiagen). Elute purified PCR products in 2 × 15 µl molecular grade dH_2O.
10. Calculate the concentration of purified PCR products using a nanodrop spectrophotometer and adjust accordingly for sequencing. For products under 200 bp, prepare a minimum of 15 µl sample at a concentration of 2 ng/µl. For PCR products 200–400 bp in length, prepare a minimum of 15 µl sample at a concentration of 5 ng/µl.
11. Sequence PCR products in the forward direction (Eurofins MWG Operon) and analyse results by comparison to the predicted bisulfite converted sequence generated from the original DNA sequence (*see* Fig. 2). Unmethylated cytosine (C) will undergo conversion to uracil (T), while methylated cytosine will remain unreactive. Visualization of bisulfite sequencing results are generally achieved by graphing filled circles to represent methylated CpG sites and unfilled circles to represent unmethylated CpG sites (*see* Fig. 2).

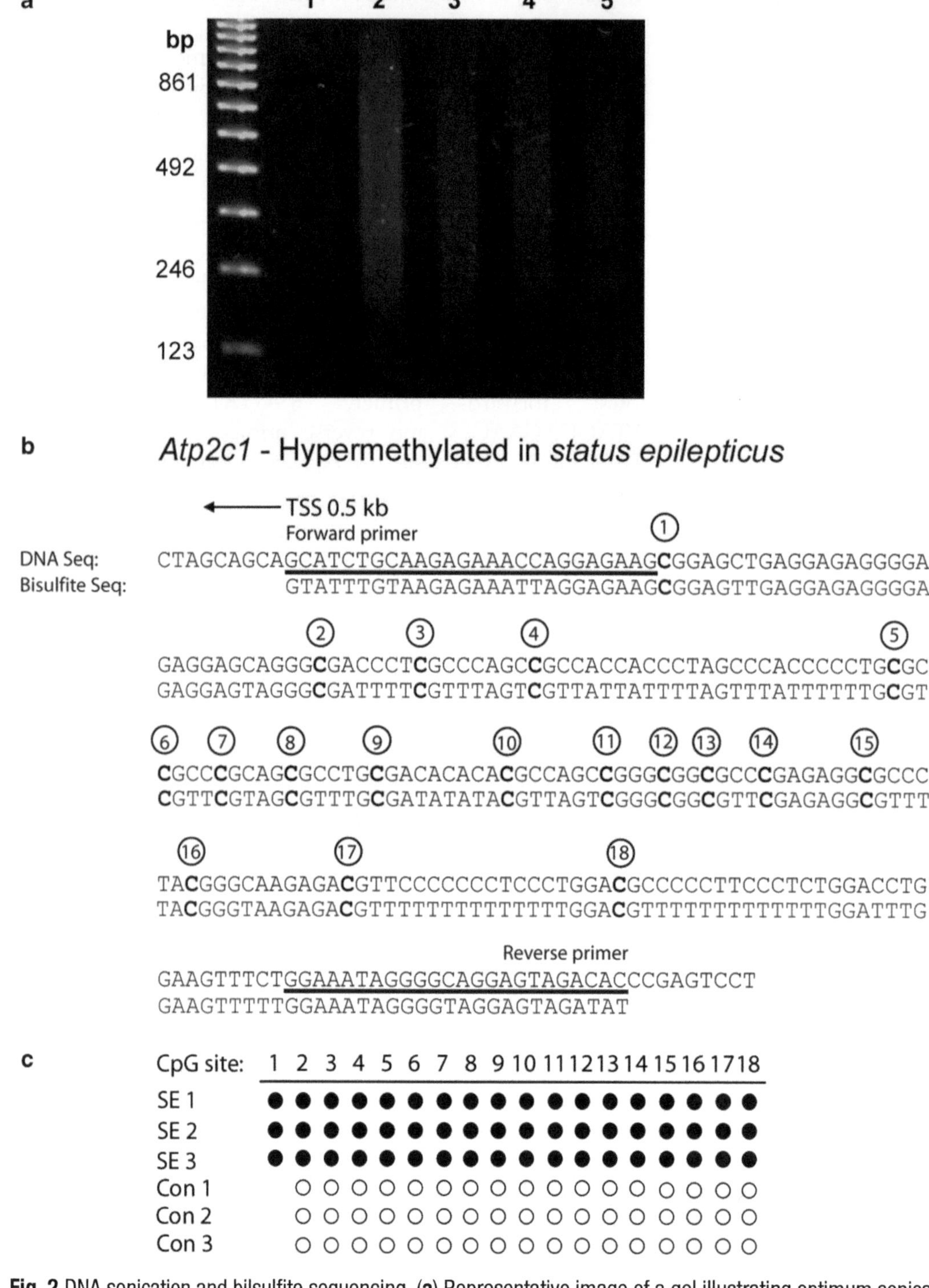

Fig. 2 DNA sonication and bilsulfite sequencing. (**a**) Representative image of a gel illustrating optimum sonication of 5 μg DNA. *Lane 1*, negative control; *Lane 2*, four sonications; *Lane 3*, five sonications; *Lane 4*, six sonications; *Lane 5* (optimum), seven sonications. Optimal sonication was adjusted based on fragment size being between 200 and 800 bp. (**b**) Schematic of bisulfite sequencing PCR design for *Atp2c1*. The regions of forward and reverse primer design are indicated, along with the distance to the transcriptional start site (TSS) of the gene and the predicted sites of cytosine methylation (numbered 1–18). Both the genomic sequence and the predicted bisulfite converted sequence are shown. (**c**) Schematic of typical bisulfite sequencing results for *Atp2c1* in this mouse model of SE. Results show that following SE, all 18 CpG sites are differentially hypermethylated (*filled circles*) across three individual biological replicates (SE1-3) when compared to control animals (Con1-3)

4 Notes

1. Custom arrays can alternatively be designed and probes should cover a tiled region of −1,300 kb to +500 bp of the promoter of genes of interest. Probes should be of 50–75 bp in length and should be spaced approximately 100 bp apart.
2. Microdissection of hippocampal tissue should be carried out on cool block as quickly as possible. Tissue should be flash frozen promptly following dissection. This is particularly critical when carrying out simultaneous extraction of protein and RNA along with DNA. Protein and RNA degradation can occur rapidly at room temperature. All dissection instruments should be cleaned well before use and between animals using ethanol and RNAse inhibitors where appropriate. This prevents cross-contamination of samples and introduction of degrading enzymes.
3. Successful simultaneous extraction of protein, RNA, and DNA has been carried out using the All-in-One Purification Kit (Norgen).
4. Depending on the extraction method used, isolated DNA may be contaminated with RNA. For this reason, it is critically important to pretreat all DNA with RNAse A. This is particularly relevant when using kits such as the All-in-One Purification Kit (Norgen) which simultaneously extract DNA, RNA, and Protein.
5. The sonicator should be cleaned well before use and a water blank sonicated between each sample in order to prevent sample cross-contamination. Minute breaks between sonications are critical for efficient sonication of DNA and to prevent overheating of samples.
6. 10× and 1× IP buffer can be stored at room temperature for a month.
7. Antibody for MeDIP should be kept in 10 μl aliquots at −20 °C for use with individual samples, as repeated freezing and thawing may affect MEDIP efficiency.
8. Labelled DNA must be stored at −80 °C and used within a month. After this time, the quality of labeling deteriorates and will have a negative impact on the quality of array data retrieved.
9. Extreme care must be taken not to introduce bubbles into samples when loading them onto arrays. Excess volume should always be included in the design of sample preparation in order to allow residual volume to be left in the sample tube thereby avoiding bubble introduction. Pipettes should be held perpendicular to the slide and with slight pressure applied to the tip, to ensure a tight seal when loading.

10. Vigorous agitation during washing steps for the full time indicated in the manufacturer's protocol is critical to efficient washing of microarrays. Inadequate agitation can lead to background fluorescence which can have dramatic effects of array results. In addition, arrays should not be allowed to dry in between washing steps.
11. Bisulfite converted DNA should be used promptly in downstream applications (pcr and sequencing) and should not be stored for long periods.

Acknowledgments

This work was supported by the Science Foundation of Ireland (Award 08/IN.1./B1875) and by the Medical Research Charities Group/Health Research Board Joint Funding Scheme through Brainwave, the Irish Epilepsy Association.

References

1. Feng J, Fan G (2009) The role of DNA methylation in the central nervous system and neuropsychiatric disorders. Int Rev Neurobiol 89:67–84
2. Robertson KD (2005) DNA methylation and human disease. Nat Rev Genet 6:597–610
3. Jaenisch R, Bird A (2003) Epigenetic regulation of gene expression: how the genome integrates intrinsic and environmental signals. Nat Genet 33(Suppl):245–254
4. Iraola-Guzman S, Estivill X, Rabionet R (2011) DNA methylation in neurodegenerative disorders: a missing link between genome and environment? Clin Genet 80:1–14
5. Miller CA, Sweatt JD (2007) Covalent modification of DNA regulates memory formation. Neuron 53:857–869
6. Miller-Delaney SF, Das S, Sano T, Jimenez-Mateos EM, Bryan K, Buckley PG et al (2012) Differential DNA methylation patterns define status epilepticus and epileptic tolerance. J Neurosci 32:1577–1588
7. Urdinguio RG, Sanchez-Mut JV, Esteller M (2009) Epigenetic mechanisms in neurological diseases: genes, syndromes, and therapies. Lancet Neurol 8:1056–1072
8. Guo JU, Ma DK, Mo H, Ball MP, Jang MH, Bonaguidi MA et al (2011) Neuronal activity modifies the DNA methylation landscape in the adult brain. Nat Neurosci 14:1345–1351
9. Levenson JM, Roth TL, Lubin FD, Miller CA, Huang IC, Desai P et al (2006) Evidence that DNA (cytosine-5) methyltransferase regulates synaptic plasticity in the hippocampus. J Biol Chem 281:15763–15773
10. Zhu Q, Wang L, Zhang Y, Zhao FH, Luo J, Xiao Z et al (2012) Increased expression of DNA methyltransferase 1 and 3a in human temporal lobe epilepsy. J Mol Neurosci 46:420–426
11. Kobow K, Jeske I, Hildebrandt M, Hauke J, Hahnen E, Buslei R et al (2009) Increased reelin promoter methylation is associated with granule cell dispersion in human temporal lobe epilepsy. J Neuropathol Exp Neurol 68: 356–364
12. Aminoff MJ, Simon RP (1980) Status epilepticus. Causes, clinical features and consequences in 98 patients. Am J Med 69:657–666
13. Fujikawa DG, Itabashi HH, Wu A, Shinmei SS (2000) Status epilepticus-induced neuronal loss in humans without systemic complications or epilepsy. Epilepsia 41:981–991
14. Hatazaki S, Bellver-Estelles C, Jimenez-Mateos EM, Meller R, Bonner C, Murphy N et al (2007) Microarray profile of seizure damage-refractory hippocampal CA3 in a mouse model of epileptic preconditioning. Neuroscience 150:467–477
15. Jimenez-Mateos EM, Hatazaki S, Johnson MB, Bellver-Estelles C, Mouri G, Bonner C et al (2008) Hippocampal transcriptome after status epilepticus in mice rendered seizure damage-tolerant by epileptic preconditioning features suppressed calcium and neuronal

excitability pathways. Neurobiol Dis 32: 442–453

16. Jimenez-Mateos EM, Henshall DC (2009) Seizure preconditioning and epileptic tolerance: models and mechanisms. Int J Physiol Pathophysiol Pharmacol 1:180–191
17. Jimenez-Mateos EM, Mouri G, Conroy RM, Henshall DC (2010) Epileptic tolerance is associated with enduring neuroprotection and uncoupling of the relationship between CA3 damage, neuropeptide Y rearrangement and spontaneous seizures following intra-amygdala kainic acid-induced status epilepticus in mice. Neuroscience 171:556–565
18. Jimenez-Mateos EM, Engel T, Merino-Serrais P, McKiernan RC, Tanaka K, Mouri G et al (2012) Silencing microRNA-134 produces neuroprotective and prolonged seizure-suppressive effects. Nat Med 18:1087–1094
19. Weber M, Davies JJ, Wittig D, Oakeley EJ, Haase M, Lam WL, Schubeler D (2005) Chromosome-wide and promoter-specific analyses identify sites of differential DNA methylation in normal and transformed human cells. Nat Genet 37:853–862
20. Buckley PG, Das S, Bryan K, Watters KM, Alcock L, Koster J et al (2011) Genome-wide DNA methylation analysis of neuroblastic tumors reveals clinically relevant epigenetic events and large-scale epigenomic alterations localized to telomeric regions. Int J Cancer 128:2296–2305
21. Das S, Bryan K, Buckley PG, Piskareva O, Bray IM, Foley N et al (2013) Modulation of neuroblastoma disease pathogenesis by an extensive network of epigenetically regulated microRNAs. Oncogene 32(24):2927–2936
22. Das S, Foley N, Bryan K, Watters KM, Bray I, Murphy DM et al (2010) MicroRNA mediates DNA demethylation events triggered by retinoic acid during neuroblastoma cell differentiation. Cancer Res 70:7874–7881
23. Paxinos G, Franklin KBJ (1997) The mouse brain in stereotaxic coordinates. Academic, San Diego, CA
24. Lein ES, Zhao X, Gage FH (2004) Defining a molecular atlas of the hippocampus using DNA microarrays and high-throughput in situ hybridization. J Neurosci 24:3879–3889

Part IV

Data Analysis

Chapter 8

Integrative Analysis of ChIP-Chip and ChIP-Seq Dataset

Lihua Julie Zhu

Abstract

Epigenetic regulation and interactions between transcription factors and regulatory genomic regions play crucial roles in controlling transcriptional regulatory networks that drive development, environmental responses, and disease. Chromatin immunoprecipitation (ChIP) followed by high-throughput sequencing (ChIP-seq) and ChIP followed by genomic tiling microarray hybridization (ChIP-chip) are the two of the most widely used technologies for genome-wide identification of DNA protein interactions and histone modification in vivo. Many algorithms and tools have been developed and evaluated that allow identification of transcription factor binding sites from ChIP-seq or ChIP-chip datasets. However, binding site identification is only the first step; the ultimate goal is to discover the regulatory network of the transcription factor (TF). Here, we present a common workflow for downstream analysis of ChIP-chip and ChIP-seq with an emphasis on annotating binding sites and integration with gene expression data to identify direct and indirect targets of the TF. These tools will help with the overall goal of unraveling transcriptional regulatory networks using datasets publicly available in GEO.

Key words ChIP-chip, ChIP-seq, Gene expression, Regulatory network, Microarray, GO enrichment analysis, Pathway analysis, Integrated analysis, Transcription factor, Motif discovery

1 Introduction

One of the most fundamental questions in biology is how the regulation of gene expression patterns is orchestrated so precisely for different cell types, different tissues, and different developmental stages, and what goes wrong during the development of various diseases. The basic building block of the regulatory network is the site where transcription factors (TF) bind to DNA, turning a gene on or off. One of the approaches to untangle the complex regulatory network of a TF is to discover the binding sites of a particular TF in the genome and determine whether TF binding affects the expression of nearby genes. Chromatin immunoprecipitation (ChIP) followed by high-throughput sequencing (ChIP-seq) and ChIP followed by genomic tiling microarray hybridization (ChIP-chip) have become standard high-throughput technologies for genome-wide identification of TF target sites and histone modification in vivo [1–4].

Tin-Lap Lee and Alfred Chun Shui Luk (eds.), *Tiling Arrays: Methods and Protocols*, Methods in Molecular Biology, vol. 1067, DOI 10.1007/978-1-62703-607-8_8, © Springer Science+Business Media New York 2013

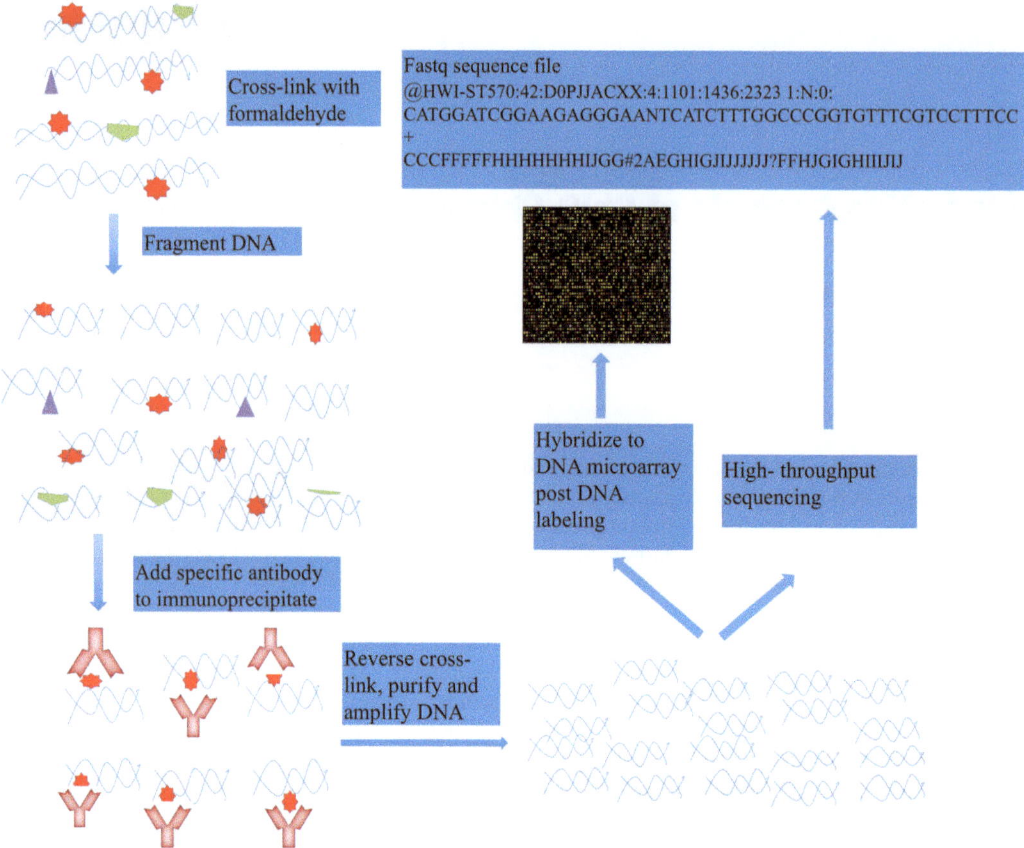

Fig. 1 Wet-lab workflow for ChIP-seq and ChIP-chip experiments. DNA and proteins are cross-linked and purified with an antibody specific to the protein of interest. Purified DNA is amplified followed by high-throughput sequencing or by hybridization to tiling Microarrays

Proteins and DNA are cross-linked with formaldehyde, followed by fragmentation via sonication and immunoprecipitation with an antibody specific to a TF. The immunoprecipitated DNA fragments are subject to sequencing or hybridization to a genomic tiling microarray (Fig. 1). Many technical considerations to obtain high-quality data are common to ChIP-seq and ChIP-chip, including antibody selection, fragment size control, amplification bias, appropriate controls, and sufficient numbers of biological replicates (reviewed in refs. 5 and 6). For a complete comparison of ChIP-chip and ChIP-seq methods, please see previous reviews [7, 8]. The effects of different tiling array platforms, amplification protocols, and signal detection algorithms have been compared using a ChIP-chip experiment with predefined DNA targets [4]. For a technical review of template preparation, sequencing and imaging, genome alignment and assembly approaches, and current NSG platforms, please see the review in ref. 9.

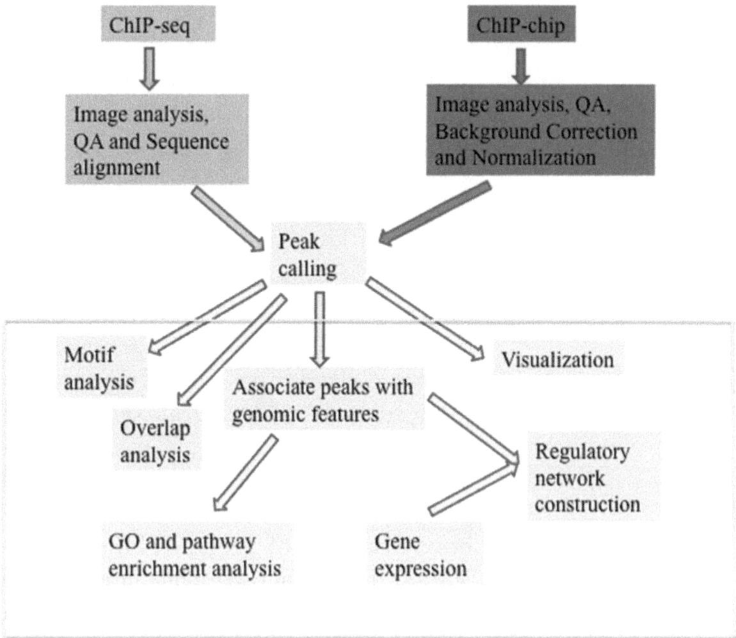

Fig. 2 Analysis workflows for ChIP-seq and ChIP-chip experiments. ChIP-seq analysis workflow and ChIP-chip workflow share a common downstream analysis workflow as depicted inside the *green box*, including associating peaks to genomic features, overlap analysis, motif discovery, GO and pathway enrichment analysis, and regulatory network construction (Color figure online)

Figure 2 depicts a typical analysis workflow for ChIP-seq and ChIP-chip. For ChIP-seq experiments, sequence information is extracted from image files with image analysis software such as Illumina Firecrest and Bustard applications. Sequences are typically saved in a FASTQ format [10]. Although several variants of FASTQ format exist differing in the quality score definition and header annotation, each sequence is represented in FASTQ file in four lines in the order of header line starting with @, sequence line, signal end of sequence line starting with +, and read base quality score (*see* example in Fig. 1). The quality score and the header annotation information are frequently used as quality assessment (QA) and read quality filters to remove low confidence sequencing reads. The remaining high-quality sequences are mapped onto the reference genome using software such as Bowtie [11], BWA [12], and SOAP [13] (reviewed in ref. 14). A peak-calling algorithm, such as MACS [15], FindPeaks [16], GeneTrack [17], QuEST [3], SISSRs [18], Useq [19], BayesPeak [20], CisGenome [21], CSAR [22], DIME [23], and PICS [24], is applied to identify the binding sites (also referred to as peaks) that are significantly enriched in the ChIP sample compared to the control sample and/or enriched in the local region compared to the immediate neighboring region(s), sometimes taking advantage of strand-specific density profiles, and/or genome-wide read mappability. In general, these peak detection algorithms can be divided into sliding window approaches such as CisGenome, MACS, and Useq; local maxima estimation approaches such as FindPeaks and PICS; and hidden Markov Model (HMM) approaches such as BayesPeak. For a detailed evaluation and comparisons among these algorithms, please see reviews in ref. 25, 26. Peak locations are usually converted to a format, such as BED or Wiggle (WIG), that can be uploaded to the UCSC Genome Browser, an open-access, web-based, up-to-date source for genome sequence data integrated with a large collection of related annotations [27]. Batch annotation tools, such as ChIPpeakAnno [28], are commonly used to associate peaks to the rich genomic annotations, such as promoter region, miRNA, exon, intron, CpG island, gene ontology, and KEGG pathway. In addition, peak sequences are frequently extracted for de novo discovery of TF DNA binding site motifs using tools such as MEME [29], DREME [30], and rGADEM [31].

Overall, a ChIP-chip data analysis workflow is very similar to that of ChIP-seq. The unique aspect of ChIP-chip data analysis involves background correction and signal intensity normalization within array and among arrays prior to peak calling (reviewed in ref. 4). The subsequent downstream analysis is exactly the same as in ChIP-seq.

Here we present a common downstream analysis workflow shared by ChIP-chip and ChIP-seq experiments. Step-by-step analysis of a publicly available dataset is illustrated using tools mainly developed in the R environment, including associating peaks with rich genomic annotations, analysis of GO and pathway enrichment, detection of *bona fide* TF binding motifs and construction of regulatory networks integrated with gene expression datasets. These downstream steps are shown enclosed in the green box in Fig. 2. Publicly available ChIP-chip data for the *C. elegans* TF DAF-12 and gene expression data for animals lacking this TF are used as an example and some recovered TF binding motifs, GO terms, and interaction pathways are consistent with the known functions of this protein.

2 Materials

ChIP-chip peaks were downloaded from GEO at http://www.ncbi.nlm.nih.gov/geo/query/acc.cgi?acc=GSE28350 [32] (*see* **Note 1**). Expression Microarray results were downloaded from [33]. Software tools are listed in Table 1 (*see* **Note 2**).

Table 1
Software tools for downstream analysis of ChIP-chip and ChIP-seq

Software	URL	Reference
R 2.15 or later	http://cran.r-project.org/	[38]
ChIPpeakAnno 2.5.8 or later	http://www.bioconductor.org/packages/release/bioc/html/ChIPpeakAnno.html	[28]
liftOver	http://genome.ucsc.edu/cgi-bin/hgLiftOver	[39]
rtracklayer	http://www.bioconductor.org/packages/release/bioc/html/rtracklayer.html	[40]
rGADEM	http://www.bioconductor.org/packages/release/bioc/html/rGADEM.html	[31]
MotIV	http://www.bioconductor.org/packages/release/bioc/html/MotIV.html	[36, 41]
STAMP	http://www.benoslab.pitt.edu/stamp/	[36]
GeneNetworkBuilder	http://pgfe.umassmed.edu/bioinformatics/GeneNetworkBuilder/ http://www.bioconductor.org/packages/release/bioc/html/GeneNetworkBuilder.html	[42]

3 Methods

3.1 Determining and Visualizing the Concordance Among the Biological Replicates

In the first step, load the *ChIPpeakAnno* package and three example datasets as bed format that contain DAF-12 binding sites identified in *C. elegans* from three biological replicates [32].

```
Library(ChIPpeakAnno)
DAF12.peaks.rep1 = BED2RangedData(read.table("/Users/zhuj/Documents/Publication/JulieZhu/BookChapter/dataset/DataForExample/GSM700907_DAF-12_TAP_vs_INPUT_#1_WS170.bed", sep="\t", skip=5))
DAF12.peaks.rep2 = BED2RangedData(read.table("/Users/zhuj/Documents/Publication/JulieZhu/BookChapter/dataset/DataForExample/GSM700908_DAF-12_TAP_vs_INPUT_#2_WS170.bed", sep="\t", skip=5))
DAF12.peaks.rep3 = BED2RangedData(read.table("/Users/zhuj/Documents/Publication/JulieZhu/BookChapter/dataset/DataForExample/GSM700909_DAF-12_TAP_vs_INPUT_#3_WS170.bed", sep="\t", skip=5))
```

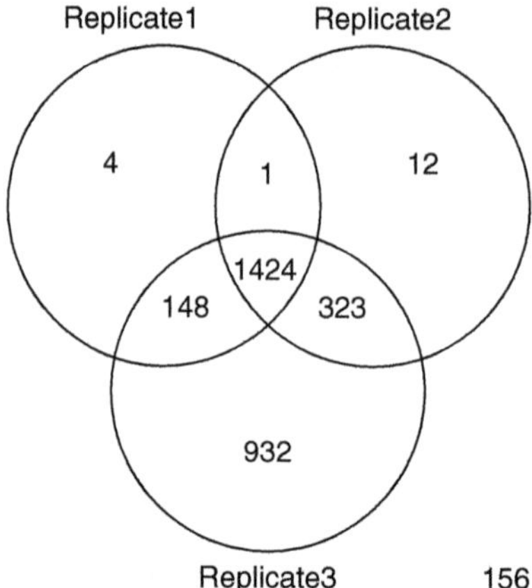

Fig. 3 Overlapping of putative DAF-12-binding sites among three biological replicates in worm. The Venn diagram was generated from the putative DAF-12-binding sites of three biological replicates in worm [32]. Hypergeometric test shows that there is significant overlapping among biological replicates (p-value < 2e-48 for all three pair-wise comparisons)

In the next step, generate a Venn diagram using the function *makeVennDiagram* to visualize the overlap among the three replicates (Fig. 3). In addition, pair-wise overlapping significance tests demonstrate significant concordance among replicates (p-value < 2e–48) (*see* **Note 3**).

```
pdf("Fig.3.pdf")
makeVennDiagram(RangedDataList(DAF12.peaks.rep1, DAF12.peaks.rep2, DAF12.peaks.rep3), NameOfPeaks=c("Replicate1","Replicate2","Replicate3"), maxgap=0, totalTest=3000)
dev.off()
```

3.2 Combining Peaks from Replicates, Filtering Peaks Present in Negative Control and Associating the Merged Peaks with Features of Interest

Prior to associating features of interest with the peaks, it is a common practice to combine peaks from replicates.

```
MergedPeaks=findOverlappingPeaks(findOverlappingPeaks(DAF12.peaks.rep1, DAF12.peaks.rep2, maxgap=0, select="first", NameOfPeaks1="R1", NameOfPeaks2="R2")$MergedPeaks, DAF12.peaks.rep3, maxgap=0, select="first", NameOfPeaks1="Replicate1Repliate2", NameOfPeaks2="R3")$MergedPeaks
```

Then any peaks also present in the DAF-12(0) negative control are filtered out.

```
negative.control = BED2RangedData(read.table("/Users/zhuj/Documents/Publication/JulieZhu/BookChapter/dataset/DataForExample/GSM700906_AA86_vs_INPUT_WS170.bed",sep="\t", skip=5))
```

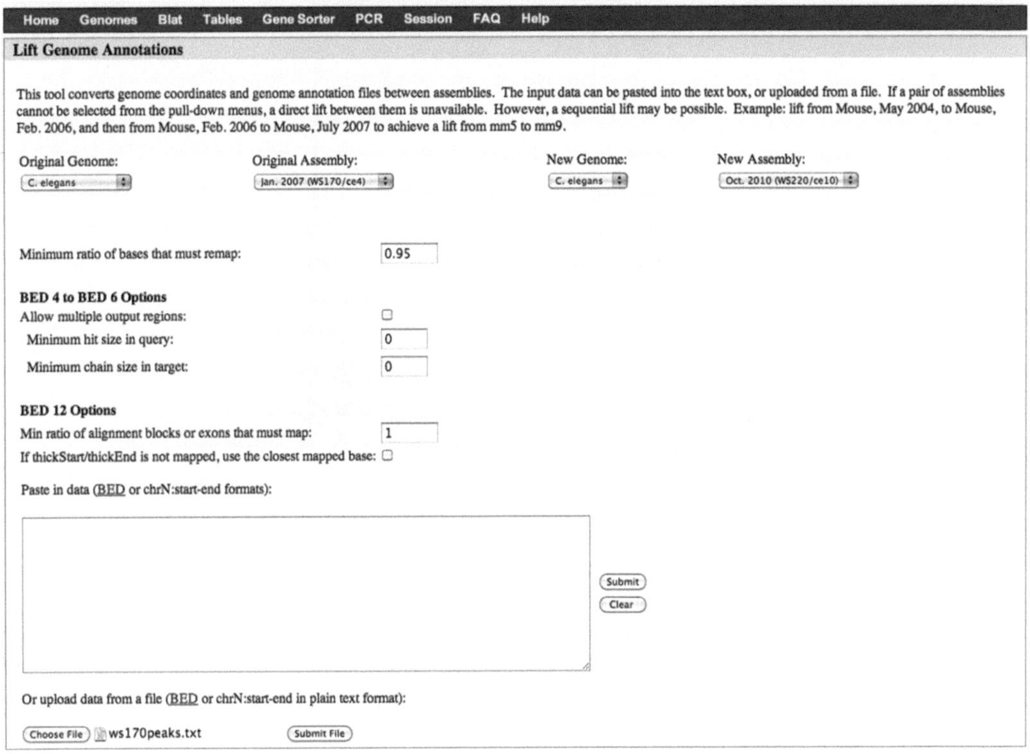

Fig. 4 A screenshot of the liftover tool in the UCSC Genome Browser. The UCSC liftover tool was used to convert genome coordinates from WS170 to WS220

```
final.peaks=MergedPeaks[!rownames(MergedPeaks) %in% rownames(findOverlappingPeaks(MergedPeaks, negative.control, select="first", maxgap=0, NameOfPeaks1="M", NameOfPeaks2="NC")$Peaks1withOverlaps),]
```

The downloaded peaks are from an old genome version WS170. However, the current genome version of *C. elegans* is WS220. To convert WS170 coordinates to corresponding WS220, save the final peaks as a file in a six-field BED format (*see* **Note 4**) and use the liftover tool in UCSC Genome Browser illustrated in Fig. 4. Alternatively, the *liftOver* function implemented in the *rtracklayer* package can be used instead.

```
temp=as.data.frame(final.peaks)[,-4]
temp[,1]=paste("chr", temp[,1],sep="")
temp[,5]=0
temp[,6]="+"
write.table(temp, file = "ws170peaks.txt",sep = "\t", row.names = FALSE, col.names=FALSE,quote=FALSE)
WS220.peaks = BED2RangedData(read.table("WS220.bed", sep="\t",header=FALSE))
```

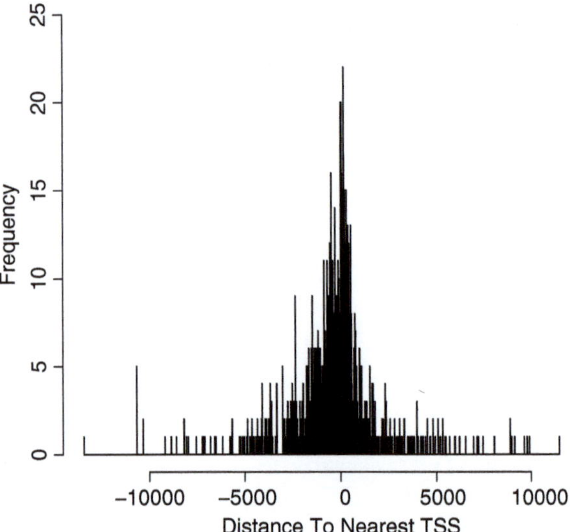

Fig. 5 Distribution of DAF-12-binding sites relative to nearest TSSs in worm. The histogram was generated from the putative DAF-12-binding regions merged from three biological replicates identified in worm [32]. The plot shows that DAF-12-binding sites are enriched in regions more symmetrically around transcription start sites. The mean of the distance to the nearest TSS is—330 ± 62 bases (mean ± SEM)

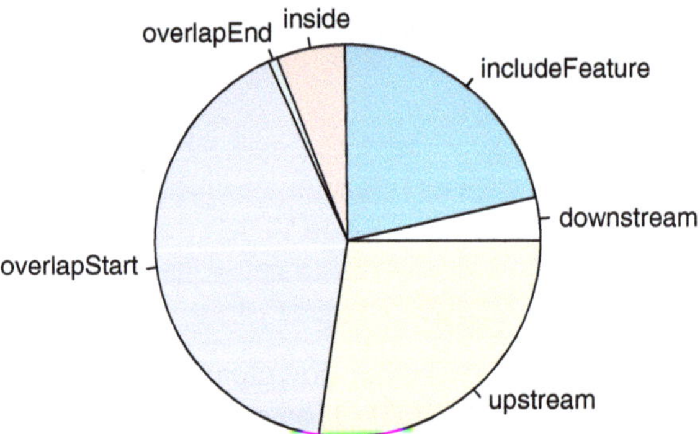

Fig. 6 Pie chart of DAF-12-binding sites relative to nearest genes in worm. The pie chart was generated from the putative DAF-12-binding regions merged from three biological replicates identified in worm [32]. The plot shows that DAF-12-binding sites are distributed more on upstream and around the genes

Associating the final peaks with the nearest gene using *annotatePeakInBatch* function reveals that DAF-12 binding sites are distributed densely around TSSs (Figs. 5 and 6).

```
mart = useMart(biomart = "ensembl", dataset = "celegans_gene_ensembl") (see Note 5)
TSS = getAnnotation(mart, featureType = "TSS") (see Note 6)
```

```
annotated.DAF12.peaks=annotatePeakInBatch(WS220.peaks, AnnotationData=TSS) (see Note 7)
```

Additional annotations such as entrez ID, gene symbol, and gene name can be added using function *addGeneIDs* and the annotated peaks can be saved as an Excel file or plotted for visualizing the peak distribution relative to the genomic features of interest.

```
library(org.Ce.eg.db)
annotated.DAF12.peaks=addGeneIDs(annotated.DAF12.peaks,"org.Ce.eg.db", c("entrez_id", "symbol","genename")) (see Note 8)
write.table(as.data.frame(annotated.DAF12.peaks), file="AnnotatedDAF-12BindingSites.xls", sep="\t", row.names=FALSE)
y = annotated.DAF12.peaks$distancetoFeature[!is.na(annotated.DAF12.peaks$distancetoFeature)]
pdf("Figure5.pdf")
hist(y, xlab="Distance To Nearest TSS", main="", breaks=1000, xlim=c(min(y)-100, max(y)+100), ylim=c(0,25))
dev.off()
pdf("Figure6.pdf")
pie(table(annotated.DAF12.peaks$insideFeature))
dev.off()
```

3.3 Finding the Peaks That Reside in Bi-directional Promoters

The analysis result shows that more than 77 % of DAF-12 peaks reside in bi-directional promoters (*see* **Note 9**). It is important to include genes associated to the peaks in both directions for subsequent direct target identification and for GO and pathway enrichment analysis.

```
BDP = peaksNearBDP(WS220.peaks, AnnotationData=TSS, MaxDistance =5000)
c(BDP$percentPeaksWithBDP, BDP$n.peaksWith-BDP, BDP$n.peaks)
all.genes = union(annotated.DAF12.peaks$feature, BDP$peaksWithBDP$feature)
```

3.4 Summarizing Peak Distributions Over Exon, Intron, Enhancer, Proximal Promoter, 5′ UTR and 3′ UTR Regions

Obtain the annotations using getAnnotation followed by calling function *assignChromosomeRegion* to summarize the relative peak distributions among different chromosome regions.

```
utr5=getAnnotation(mart, featureType="5utr")
utr3=getAnnotation(mart, featureType="3utr")
exon=getAnnotation(mart, featureType="Exon")
Feature.distribution=assignChromosomeRegion(WS220.peaks, exon, TSS, utr5, utr3)
```

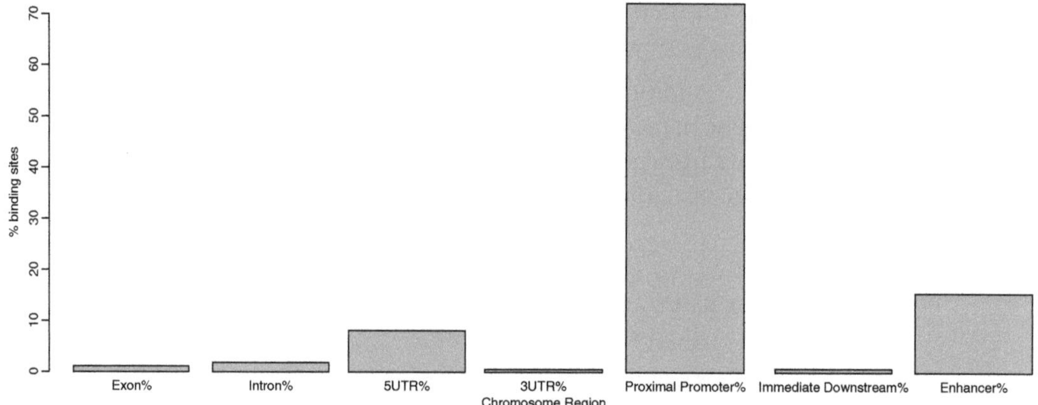

Fig. 7 Distribution of DAF-12 binding sites among various features in worm. The bar plot was generated from the putative DAF-12-binding regions merged from three biological replicates identified in worm [32]. The plot shows that 70 % DAF-12-binding sites reside in proximal promoter regions

The summarized distribution can be plotted as a bar graph and saved in pdf format.

```
pdf("Figure7.pdf")
barplot(unlist(Feature.distribution),cex.names=1, xlab="Chromosome Region", ylab="% binding sites", main="")
dev.off()
```

Figure 7 shows that 70 % DAF-12 binding sites reside in proximal promoter regions (*see* **Note 10**).

3.5 Obtaining Enriched GO Terms for Genes Near the Peaks

Load the organism-specific GO gene mapping package before acquiring enriched GO terms using *getEnrichedGO* (*see* **Note 11**).

```
library(org.Ce.eg.db)
enrichedGO<- getEnrichedGO (all.genes, orgAnn="org.Ce.eg.db", maxP=0.05, multiAdj=TRUE, minGOterm=10, multiAdjMethod="BH") (see Note 8 and 11)
write.table(enrichedGO$bp, file="Table2_EnrichedBiologicalProcesses.xls", sep="\t", row.names=FALSE)
write.table(enrichedGO$mf, file="Table2_EnrichedMolecularFunctions.xls", sep="\t", row.names=FALSE)
write.table(enrichedGO$cc, file="Table2_EnrichedCellularComponents.xls", sep="\t", row.names=FALSE)
```

Table 2 shows a list of enriched GO molecular functions and biological processes for DAF-12 binding sites (*see* **Note 11**). Note

Table 2
Enriched GO molecular functions and biological processes of DAF-12-fbinding sites in worm ordered by false discovery rate (FDR)

GO ID	GO term	GO definition	Category	FDR
GO:0005515	Protein binding	Interacting selectively and non-covalently with any protein or protein complex (a complex of two or more proteins that may include other nonprotein molecules)	MF	4.85E-08
GO:0003735	Structural constituent of ribosome	The action of a molecule that contributes to the structural integrity of the ribosome	MF	0.0002
GO:0002119	Nematode larval development	The process whose specific outcome is the progression of the nematode larva over time, from its formation to the mature structure. Nematode larval development begins with the newly hatched first-stage larva (L1) and ends with the end of the last larval stage (for example the fourth larval stage (L4) in *C. elegans*). Each stage of nematode larval development is characterized by proliferation of specific cell lineages and an increase in body size without alteration of the basic body plan. Nematode larval stages are separated by molts in which each stage-specific exoskeleton, or cuticle, is shed and replaced anew	BP	0.0013
GO:0002164	Larval development	The process whose specific outcome is the progression of the larva over time, from its formation to the mature structure. The larva is the early, immature form of an animal that at birth or hatching is fundamentally unlike its parent and must metamorphose before assuming the adult characters	BP	0.0013
GO:0009791	Post-embryonic development	The process whose specific outcome is the progression of the organism over time, from the completion of embryonic development to the mature structure. See embryonic development	BP	0.0013
GO:0015078	Hydrogen ion transmembrane transporter activity	Catalysis of the transfer of hydrogen ions from one side of a membrane to the other	MF	0.0029
GO:0040034	Regulation of development, heterochronic	Any process that modulates the rate, frequency, or extent of the progression an integrated living unit: an anatomical structure (which may be a subcellular structure, cell, tissue, or organ), or organism over time from an initial condition to a later condition at a consistent predetermined time point during its development	BP	0.039

(continued)

Table 2
(continued)

GO ID	GO term	GO definition	Category	FDR
GO:0000166	Nucleotide binding	Interacting selectively and non-covalently with a nucleotide, any compound consisting of a nucleoside that is esterified with (ortho)phosphate or an oligophosphate at any hydroxyl group on the ribose or deoxyribose	MF	0.049
GO:0005488	Binding	The selective, non-covalent, often stoichiometric, interaction of a molecule with one or more specific sites on another molecule	MF	0.049
GO:0017076	Purine nucleotide binding	Interacting selectively and non-covalently with purine nucleotides, any compound consisting of a purine nucleoside esterified with (ortho)phosphate	MF	0.049
GO:0032553	Ribonucleotide binding	Interacting selectively and non-covalently with a ribonucleotide, any compound consisting of a ribonucleoside that is esterified with (ortho)phosphate or an oligophosphate at any hydroxyl group on the ribose moiety	MF	0.049
GO:0032555	Purine ribonucleotide binding	Interacting selectively and non-covalently with a purine ribonucleotide, any compound consisting of a purine ribonucleoside that is esterified with (ortho) phosphate or an oligophosphate at any hydroxyl group on the ribose moiety	MF	0.049
GO:0035091	Phosphatidylinositol binding	Interacting selectively and non-covalently with any inositol-containing glycerophospholipid, i.e. phosphatidylinositol (PtdIns) and its phosphorylated derivatives	MF	0.049
GO:0035639	Purine ribonucleoside triphosphate binding	Interacting selectively and non-covalently with a purine ribonucleoside triphosphate, a compound consisting of a purine base linked to a ribose sugar esterified with triphosphate on the sugar	MF	0.049
GO:0051015	Actin filament binding	Interacting selectively and non-covalently with an actin filament, also known as F-actin, a helical filamentous polymer of globular G-actin subunits	MF	0.049

The enriched GO terms were obtained including genes associated to the peaks in both directions within 5 kb to the TSS. The parameter used to generate the list is *maxP*= 0.05, *multipAdj*= TRUE, *minGOterm*= 10, and *multiAdjMethod*= "BH"

Table 3
Enriched pathways in reactome database for DAF-12 binding sites in worm ordered by FDR

Pathway ID	Pathway definition	FDR
1626134	*Caenorhabditis elegans*: Regulation of gene expression in beta cells	0.004
1626136	*Caenorhabditis elegans*: Diabetes pathways	0.004
1625991	*Caenorhabditis elegans*: Peptide chain elongation	0.004
1625992	*Caenorhabditis elegans*: Eukaryotic Translation Elongation	0.004
1625772	*Caenorhabditis elegans*: GTP hydrolysis and joining of the 60S ribosomal subunit	0.006
1626131	*Caenorhabditis elegans*: Regulation of beta-cell development	0.006
1625983	*Caenorhabditis elegans*: Eukaryotic Translation Termination	0.007
1626135	*Caenorhabditis elegans*: Insulin Synthesis and Processing	0.023
1625880	*Caenorhabditis elegans*: Developmental Biology	0.032
1625773	*Caenorhabditis elegans*: Formation of a pool of free 40S subunits	0.036

The enriched pathway terms were obtained including genes associated to the peaks in both directions within 5 kb to the TSS. The parameter used to generate the list is *maxP* = 0.05, *multipAdj* = TRUE, *minGOterm* = 10, and *multiAdjMethod* = "BH"

that several of the terms are associated with development, consistent with the known role of DAF-12 in regulation of dauer formation in *C. elegans*.

3.6 Obtaining Enriched Pathway Terms Near the Peaks

Load the pathway database Reactome before acquiring enriched pathway terms using *getEnrichedPATH*.

```
library(reactome.db)
enrichedPATH = getEnrichedPATH(all.genes, orgAnn="org.Ce.eg.db", pathAnn="reactome.db", maxP=0.05, minPATHterm=10, multiAdjMethod="BH")
```
(*see* **Note 12**)
```
write.table(enrichedPATH, file = "Table3_EnrichedPATHsWithGeneIDs.xls", sep="\t", row.names=FALSE)
```

Table 3 shows a list of enriched pathways from the Reactome database for DAF-12 binding sites (*see* **Note 12**). Note that the enriched insulin and diabetes pathways are associated with *C. elegans* dauer formation, consistent with the known role of DAF-12 in this process.

3.7 Retrieving the Sequences Around the Binding Sites for PCR Amplification or Motif Discovery

First, load the organism-specific *BSgenome* package followed fetching sequences using *getAllPeakSequence*.

```
library(BSgenome.Celegans.UCSC.ce10)
```
(*see* **Note 13**).

```
peaksWithSequences=getAllPeakSequence(WS220.
peaks, upstream=0,
   downstream=0, genome=Celegans)
   write2FASTA(peaksWithSequences,    file="DAF-
12peaks.fa", width=50)
```

Next, load *rGADEM library* for de nova motif discovery.

```
library(rGADEM)
sequences<- read.DNAStringSet("DAF-12peaks.fa",
"fasta")
gadem<-GADEM(sequences,verbose=1,genome=Cel
egans)
consensus(gadem)
motifs<- getPWM(gadem)
motifs.trimed<- trimPWMedge(motifs, threshold=1)
```

Last, load *MotIV* library for matching discovered motifs with known motifs.

```
library(MotIV) (see Note 2)
DAF12.analysis.jaspar<-  motifMatch(motifs.
trimed)
summary(DAF-12DAF12.analysis.jaspar)
pdf("Figure8.pdf")
plot(DAF12.analysis.jaspar,ncol=2,
rev=FALSE, main="Logo", bysim=TRUE)
dev.off()
```

Significant motifs (p-value < 0.0002) identified by GADEM with matching motifs in JASPAR database [34] are given in Fig. 8 as sequence logos, a graphical representation of sequence alignments [35] (*see* **Note 14**). An alternative web tool for motif matching is STAMP [36]. Note that DAF-12 is a member of the nuclear hormone receptor family of TFs and cooperates with DAF-16, a member of the forkhead family of TFs; our analysis recovered motifs for a mammalian hormone receptor, NR2F1, and a forkhead TF, FOXD3.

3.8 Constructing Regulatory Network Combining with Expression Dataset

Intersecting genes with promoter regions bound by TAF-12 from the ChIP experiment and genes differentially expressed in TAF-12 mutants allow us to identify genes directly regulated by TAF-12. Genes differentially expressed in TAF-12 mutant but not bound by TAF-12 are predicted indirect targets of TAF-12. Figure 9, generated using GeneNetworkBuilder [42], gives a system-level view of the intertwined connections among the direct and indirect targets of DAF-12, which shows that DAF-12 is a master regulator.

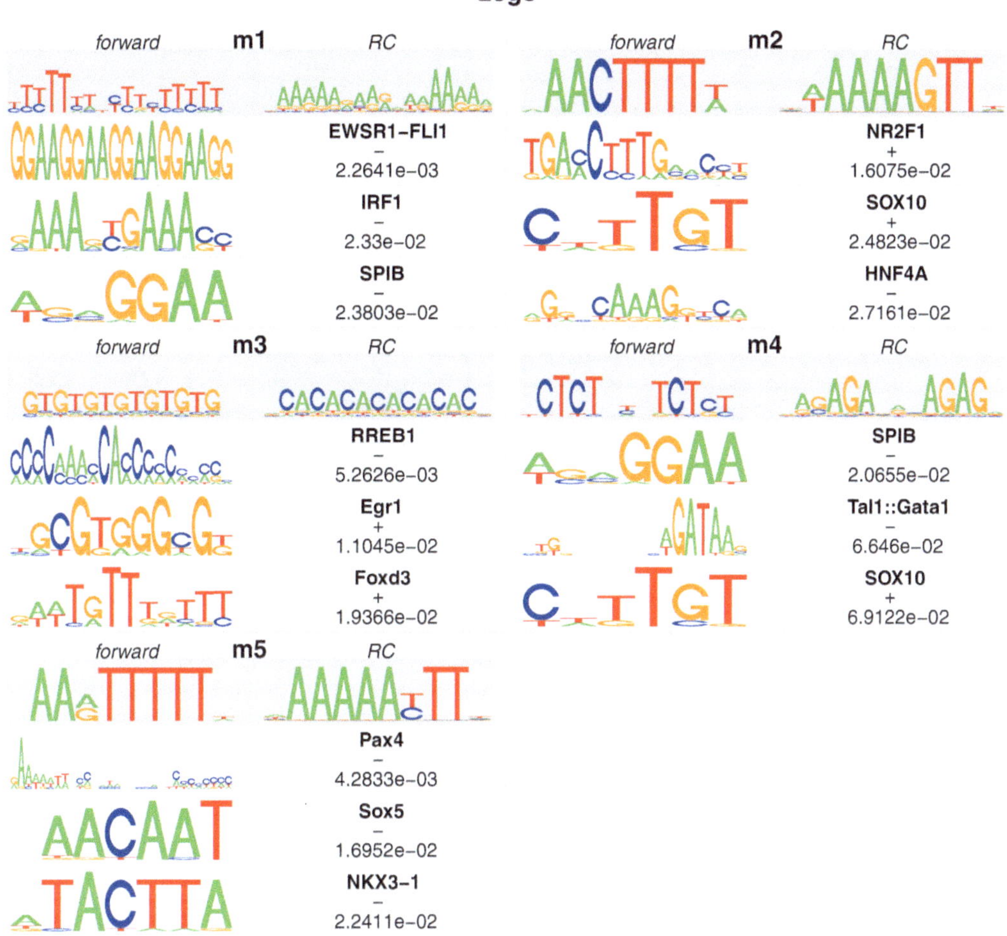

Fig. 8 Identified motifs of DAF-12-binding sites in worm and corresponding matching motifs in JASPAR database. The motifs were generated using rGADEM and the sequences from the putative DAF-12-binding regions merged from three biological replicates identified in worm [32]. The matching motifs in JASPAR database were obtained using MotIV. Five significant motifs are shown. For each discovered motif, both forward and reverse complement logos are given on the *top* and three significant matching known motifs are given below with forward motif logo, motif name, and matching *p*-value

4 Notes

1. DAF-12 binding sites from three ChIP-chip datasets were used as example. The same analysis would apply to ChIP-seq datasets as long as binding sites contain chromosome coordinates.
2. The GNU Scientific Library (GSL) needs to be installed for the MotIV package. GSL is free and can be downloaded at http://www.gnu.org/ software/gsl/for Unix distributions and at http://gnuwin32.sourceforge.net/packages/gsl.htm for Windows. ChIPpeakAnno, rGADEM, MotIV, and rtrack-

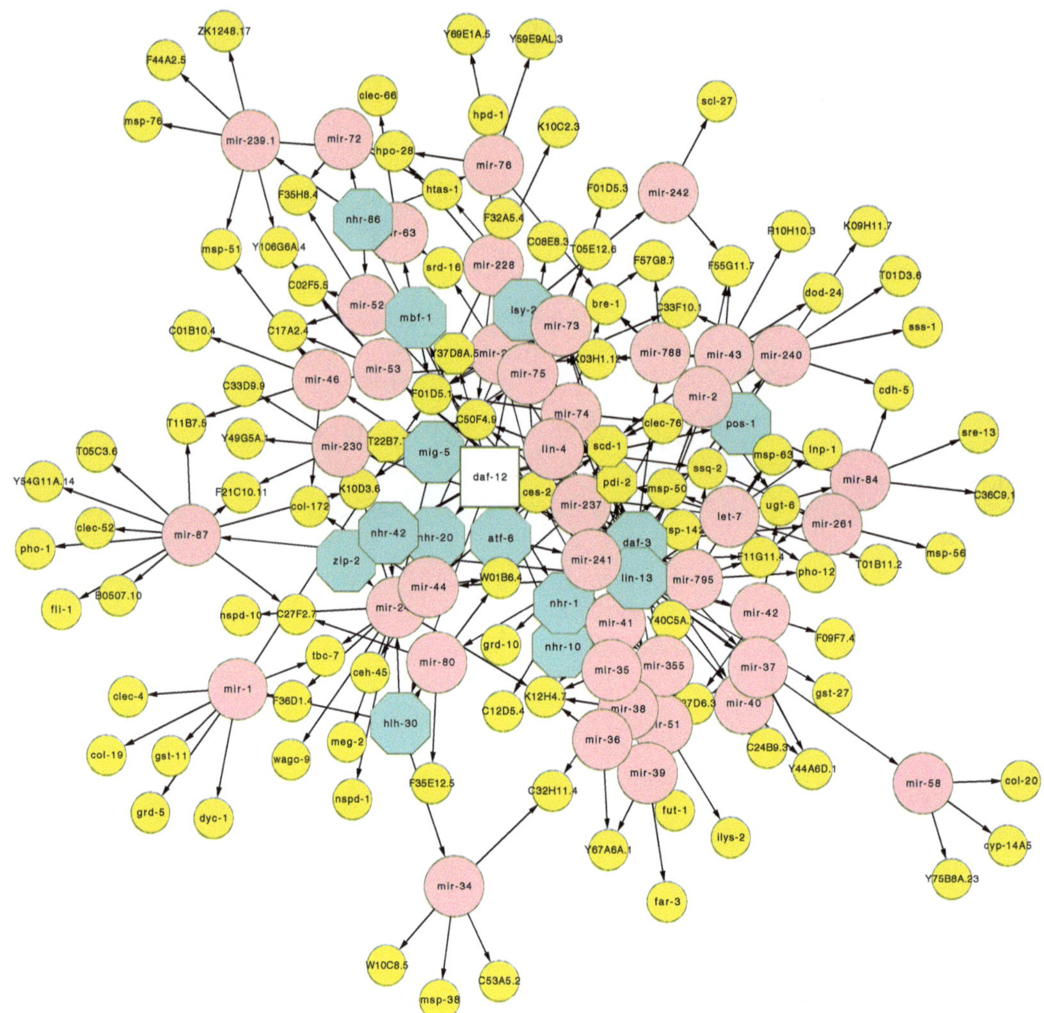

Fig. 9 Identified transcription regulatory network of DAF-12 in worm. The network was generated using GeneNetworkBuilder with the data from the putative DAF-12-binding regions merged from three biological replicates identified in worm [32] and expression microarray results from animals lacking DAF-12 [33]. Please note that it would be ideal if the worm stages are similar between the ChIP experiment and expression experiment. For this reason, the tolerance is set to two unverified internal nodes. DAF-12, the root node, is shown as a *square*, the direct targets are depicted as octagons, and all the other nodes are shown as *circles*. The larger the *circle* or *octagon*, the more downstream connections it has. The genes with altered expressions are colored in *yellow*. The genes with binding sites near the promoter region but with unchanged expression are colored in *cyan*. All the miRNAs are colored in *pink*

layer are available from Bioconductor, an open source and open development software project specializing in biological data analysis and integration based on R, a system for statistical computation and graphics [37]. R can be obtained at http://cran.r-project.org/. The bioconductor packages can be installed from R by typing *source("http://bioconductor.org/biocLite.R")* and *biocLite(c("ChIPpeakAnno","rtracklayer","rGADEM", "MotIV"))*.

3. The parameter *totalTest* indicates how many potential peaks in total are used in the hypergeometric test. It should be larger than the largest number of peaks in the replicates. The smaller it is set, the more stringent the test is. The time used to calculate *p*-value does not depend on the *totalTest*. For practical guidance on how to choose *totalTest*, please refer to the post at https://stat.ethz.ch/pipermail/bioconductor/2010-November/036540.html
4. BED format is a common file format that provides a flexible way to define peaks or annotations as data lines detailed at http://genome.ucsc.edu/FAQ/FAQformat#format1. For six-field BED format, each line contains six tab-separated fields in the order of chromosome name, chromosome start, chromosome end, line name, score and strand in + or –.
5. For a list of annotation dataset available for other species, use function *listDatasets*.
6. To annotate the peaks with other genomic features, use *getAnnotation* with different *featureType* such as *exon*, *5utr*, and *3utr*. It is also possible to pass customized annotation data into the function *annotatePeakInBatch*, such as a list of binding sites for the same protein from the literature, or a list of binding sites of a different protein functioning as transcription complex together with the protein under study. Prior to calling the function *annotatePeakInBatch*, it is necessary to represent both datasets as *RangedData*, where *start* is the start of the binding site, *end* is the end of the binding site, *names* is the name of the binding site, and *space* and *strand* are the chromosome name and strand, respectively, where the binding site is located.
7. By default, the distance is calculated as the distance between the start of the binding site and the TSS, which is the gene start for genes located on the forward strand and the gene end for genes located on the reverse strand. Users have the flexibility to set *PeakLocForDistance* to start, *middle or end*, and set *FeatureLocForDistance* to TSS, start, *middle*, *end* or *geneEnd*. Also, by default, the function outputs only nearest genes. If interested in including genes overlapping with the peaks in addition to the nearest genes, set *output = "both"*, *maxgap =0 and select = "all"*. For customized settings, please type help (*annotatePeakInBatch*) in R.
8. Package org.Ce.eg.db needs to be installed in R by typing source("http://bioconductor.org/biocLite.R") and biocLite("org.Ce.eg.db").

In addition, not all species-specific annotation packages are Entrez ID centric. For example, org.At.tair.db for Arabidopsis is TAIR ID centric and org.Sc.sgd.db for Yeast is orf centric. As a workaround, set *feature_type_id* as *entrez_id* for Arabindopsis and Yeast when using *addGeneIDs or getEnrichedGO.*

9. *MaxDistance*, 5,000 bp by default, specifies the maximum gap allowed between the peak start and TSS. In the example, it is set to 1 kb to be more stringent. For additional customized parameter setting, please type help (*peaksNearBDP*) in R.

10. Only available with ChIPpeakAnno 2.5.9 or later. By default, peaks that reside in within 1 kb upstream *(proximal.promoter.cutoff)* from or overlap with transcription start site are classified as proximal promoters. Peaks that reside within 1,000 bp downstream *(immediate.downstream.cutoff)* of gene end but not overlap 3′ UTR are classified as immediate downstream. Peaks that reside over 1,000 bp from gene start or gene end are classified as enhancers. Users have the flexibility to set both parameters different from default value 1,000 in the function *assignChromosomeRegion*.

11. For other organism-specific GO gene mapping packages org.xx.eg.db, please refer to http://www.bioconductor.org/packages/release/data/annotation/. The parameter *maxP* is the maximum *p*-value required to be considered significant, *multiAdj* indicates whether to apply multiple hypothesis testing adjustment, *minGOterm* is the minimum count in a genome for a GO term to be included, and *multiAdjMethod* is the multiple testing procedure to be applied. For customized setting, please type in R *help(getEnrichedGO)*. The output contains the corresponding gene entrezIDs for each enriched GO term. Table 2 is simplified with EntrezIDs removed using the following R code.

```
write.table(rbind(unique(enrichedGO$mf[,1:10]),
unique(enrichedGO$bp[,1:10])), file="Table2.
xls",sep="\t", row.names=FALSE)
```

12. The function getEnrichedPATH is only available with ChIPpeakAnno 2.4.0 or later. Many parameters are the same as for getErichedGO (*see* **Note 10**). For customized settings, please type in R help(getEnrichedPATH). The output contains the corresponding gene entrezIDs for each enriched pathway term. Table 3 is simplified with EntrezIDs removed using the following R code.

```
write.table(unique(enrichedPATH[,-2]),
file="Table3.xls",sep="\t", row.names=FALSE)
```

13. Only available in Bioconductor 2.11 or later. To install, type the following two lines in an R session.

```
source("http://bioconductor.org/biocLite.R")
biocLite("BSgenome.Celegans.UCSC.ce10")
```

14. As of today, JASPAR contains very limited motifs for worms. There is an effort going on spearheaded by Bioconductor core team member Paul Shannon to consolidate a richer motif data resource in the R environment.

Acknowledgment

I would like to thank Dr. Michael Brodsky at Program in Gene Function and Expression in University of Massachusetts Medical School for his critical review of the manuscript and his excellent suggestions.

References

1. Johnson DS, Mortazavi A, Myers RM, Wold B (2007) Genome-wide mapping of in vivo protein-DNA interactions. Science 316: 1497–1502
2. Robertson G, Hirst M, Bainbridge M, Bilenky M, Zhao Y, Zeng T et al (2007) Genome-wide profiles of STAT1 DNA association using chromatin immunoprecipitation and massively parallel sequencing. Nat Methods 4:651–657
3. Valouev A, Johnson DS, Sundquist A, Medina C, Anton E, Batzoglou S et al (2008) Genome-wide analysis of transcription factor binding sites based on ChIP-Seq data. Nat Methods 5:829–834
4. Johnson DS, Li W, Gordon DB, Bhattacharjee A, Curry B, Ghosh J et al (2008) Systematic evaluation of variability in ChIP-chip experiments using predefined DNA targets. Genome Res 18:393–403
5. Kidder BL, Hu G, Zhao K (2011) ChIP-Seq: technical considerations for obtaining high-quality data. Nat Immunol 12:918–922
6. Buck MJ, Lieb JD (2004) ChIP-chip: considerations for the design, analysis, and application of genome-wide chromatin immunoprecipitation experiments. Genomics 83:349–360
7. Park PJ (2009) ChIP-seq: advantages and challenges of a maturing technology. Nat Rev Genet 10:669–680
8. Ho JW, Bishop E, Karchenko PV, Negre N, White KP, Park PJ (2011) ChIP-chip versus ChIP-seq: lessons for experimental design and data analysis. BMC Genomics 12:134
9. Metzker ML (2010) Sequencing technologies—the next generation. Nat Rev Genet 11:31–46
10. Cock PJ, Fields CJ, Goto N, Heuer ML, Rice PM (2010) The Sanger FASTQ file format for sequences with quality scores, and the Solexa/Illumina FASTQ variants. Nucleic Acids Res 38:1767–1771
11. Langmead B, Trapnell C, Pop M, Salzberg SL (2009) Ultrafast and memory-efficient alignment of short DNA sequences to the human genome. Genome Biol 10:R25
12. Li H, Durbin R (2009) Fast and accurate short read alignment with Burrows-Wheeler transform. Bioinformatics 25:1754–1760
13. Li R, Li Y, Kristiansen K, Wang J (2008) SOAP: short oligonucleotide alignment program. Bioinformatics 24:713–714
14. Li H, Homer N (2010) A survey of sequence alignment algorithms for next-generation sequencing. Brief Bioinform 11:473–483
15. Zhang Y, Liu T, Meyer CA, Eeckhoute J, Johnson DS, Bernstein BE et al (2008) Model-based analysis of ChIP-Seq (MACS). Genome Biol 9:R137
16. Fejes AP, Robertson G, Bilenky M, Varhol R, Bainbridge M, Jones SJ (2008) FindPeaks 3.1: a tool for identifying areas of enrichment from massively parallel short-read sequencing technology. Bioinformatics 24: 1729–1730
17. Albert I, Wachi S, Jiang C, Pugh BF (2008) GeneTrack – a genomic data processing and visualization framework. Bioinformatics 24: 1305–1306
18. Jothi R, Cuddapah S, Barski A, Cui K, Zhao K (2008) Genome-wide identification of in vivo protein-DNA binding sites from ChIP-Seq data. Nucleic Acids Res 36:5221–5231

19. Nix DA, Courdy SJ, Boucher KM (2008) Empirical methods for controlling false positives and estimating confidence in ChIP-Seq peaks. BMC Bioinformatics 9:523

20. Spyrou C, Stark R, Lynch AG, Tavare S (2009) BayesPeak: Bayesian analysis of ChIP-seq data. BMC Bioinformatics 10:299

21. Ji H, Jiang H, Ma W, Wong WH (2011) Using CisGenome to analyze ChIP-chip and ChIP-seq data. Curr Protoc Bioinformatics Chapter 2:Unit2 13

22. Muino JM, Kaufmann K, van Ham RC, Angenent GC, Krajewski P (2011) ChIP-seq Analysis in R (CSAR): an R package for the statistical detection of protein-bound genomic regions. Plant Methods 7:11

23. Taslim C, Huang T, Lin S (2011) DIME: R-package for identifying differential ChIP-seq based on an ensemble of mixture models. Bioinformatics 27:1569–1570

24. Zhang X, Robertson G, Krzywinski M, Ning K, Droit A, Jones S, Gottardo R (2011) PICS: probabilistic inference for ChIP-seq. Biometrics 67:151–163

25. Wilbanks EG, Facciotti MT (2010) Evaluation of algorithm performance in ChIP-seq peak detection. PLoS One 5:e11471

26. Laajala TD, Raghav S, Tuomela S, Lahesmaa R, Aittokallio T, Elo LL (2009) A practical comparison of methods for detecting transcription factor binding sites in ChIP-seq experiments. BMC Genomics 10:618

27. Kasprzyk A, Keefe D, Smedley D, London D, Spooner W, Melsopp C et al (2004) EnsMart: a generic system for fast and flexible access to biological data. Genome Res 14:160–169

28. Zhu LJ, Gazin C, Lawson ND, Pages H, Lin SM, Lapointe DS, Green MR (2010) ChIPpeakAnno: a bioconductor package to annotate ChIP-seq and ChIP-chip data. BMC Bioinformatics 11:237

29. Bailey TL, Elkan C (1994) Fitting a mixture model by expectation maximization to discover motifs in biopolymers. Proc Int Conf Intell Syst Mol Biol 2:28–36

30. Bailey TL (2011) DREME: motif discovery in transcription factor ChIP-seq data. Bioinformatics 27:1653–1659

31. Li L (2009) GADEM: a genetic algorithm guided formation of spaced dyads coupled with an EM algorithm for motif discovery. J Comput Biol 16:317–329

32. Hochbaum D, Zhang Y, Stuckenholz C, Labhart P, Alexiadis V, Martin R et al (2011) DAF-12 regulates a connected network of genes to ensure robust developmental decisions. PLoS Genet 7:e1002179

33. Fisher AL, Lithgow GJ (2006) The nuclear hormone receptor DAF-12 has opposing effects on *Caenorhabditis elegans* lifespan and regulates genes repressed in multiple long-lived worms. Aging Cell 5:127–138

34. Sandelin A, Alkema W, Engstrom P, Wasserman WW, Lenhard B (2004) JASPAR: an open-access database for eukaryotic transcription factor binding profiles. Nucleic Acids Res 32(Database issue):D91–D94

35. Schneider TD, Stephens RM (1990) Sequence logos: a new way to display consensus sequences. Nucleic Acids Res 18: 6097–6100

36. Mahony S, Benos PV (2007) STAMP: a web tool for exploring DNA-binding motif similarities. Nucleic Acids Res 35(Web Server issue):W253–W258

37. Gentleman RC, Carey VJ, Bates DM, Bolstad B, Dettling M, Dudoit S et al (2004) Bioconductor: open software development for computational biology and bioinformatics. Genome Biol 5:R80

38. Ihaka R, Gentlemen R (1996) R: a language for data analysis and graphics. J Comput Graph Stat 5:299–314

39. Hinrichs AS, Karolchik D, Baertsch R, Barber GP, Bejerano G et al (2006) The UCSC Genome Browser Database: update 2006. Nucleic Acids Res 34(Database issue):D590–D598

40. Lawrence M, Gentleman R, Carey V (2009) rtracklayer: an R package for interfacing with genome browsers. Bioinformatics 25:1841–1842

41. Mahony S, Auron PE, Benos PV (2007) DNA familial binding profiles made easy: comparison of various motif alignment and clustering strategies. PLoS Comput Biol 3:e61

42. Ou J, Zhu LJ (2013) http://www.bioconductor.org/packages/release/bioc/html/GeneNetworkBuilder.html

Chapter 9

HAT: A Novel Statistical Approach to Discover Functional Regions in the Genome

Erdogan Taskesen, Bas Wouters, and Ruud Delwel

Abstract

Tiling arrays are useful for exploring local functions of regions of the genome in an unbiased fashion. The exact determination of those genomic regions based on tiling-array data, e.g., generated by means of hybridization with immunoprecipated DNA-fragments to the arrays is a challenge. Many different statistical methodologies have been developed to find biological relevant regions-of-interest (ROI) by using the quantitative signal intensity of each probe. We previously developed a method called Hypergeometric Analysis of Tiling arrays (HAT) for the analysis of tiling-array data, but it is developed such that it can also be used to study data derived by genome-wide deep sequencing approaches. Here we applied HAT to analyze two publicly available tiling-array data sets. After the detection of statistically significant ROI, these are often used in additional analysis for hypothesis testing. We therefore discuss, by using the results of the tiling-array experiment, pathway and motif analyses.

Key words HAT, Tiling array, Peak calling, Motif analysis, CEBPA

1 Introduction

Tiling arrays are a subtype of microarrays which are designed with probes that cover contiguous regions of a genome. The locations of probes do not necessarily cover genomic regions that are known to be functional, as is the case for gene expression or promoter arrays. Therefore tiling arrays differ from these microarrays as they are not by definition designed to cover known or predicted genes in the genome. Moreover, the coverage of probes in unknown genomic regions has been useful for exploring the genome in an unbiased fashion. Examples of applications for tiling arrays are (1) protein–DNA interaction by conducting chromatin immunoprecipitation (ChIP-on-chip) experiments [1], (2) epigenetic modifications by Methyl-DNA immunoprecipitation [2] (MeDIP-on-chip), or (3) identification of DNAse hypersensitive sites, which can be used to predict regulatory elements such as promoter regions, enhancers, and silencers [3]. Although tiling arrays are

Tin-Lap Lee and Alfred Chun Shui Luk (eds.), *Tiling Arrays: Methods and Protocols*, Methods in Molecular Biology, vol. 1067, DOI 10.1007/978-1-62703-607-8_9, © Springer Science+Business Media New York 2013

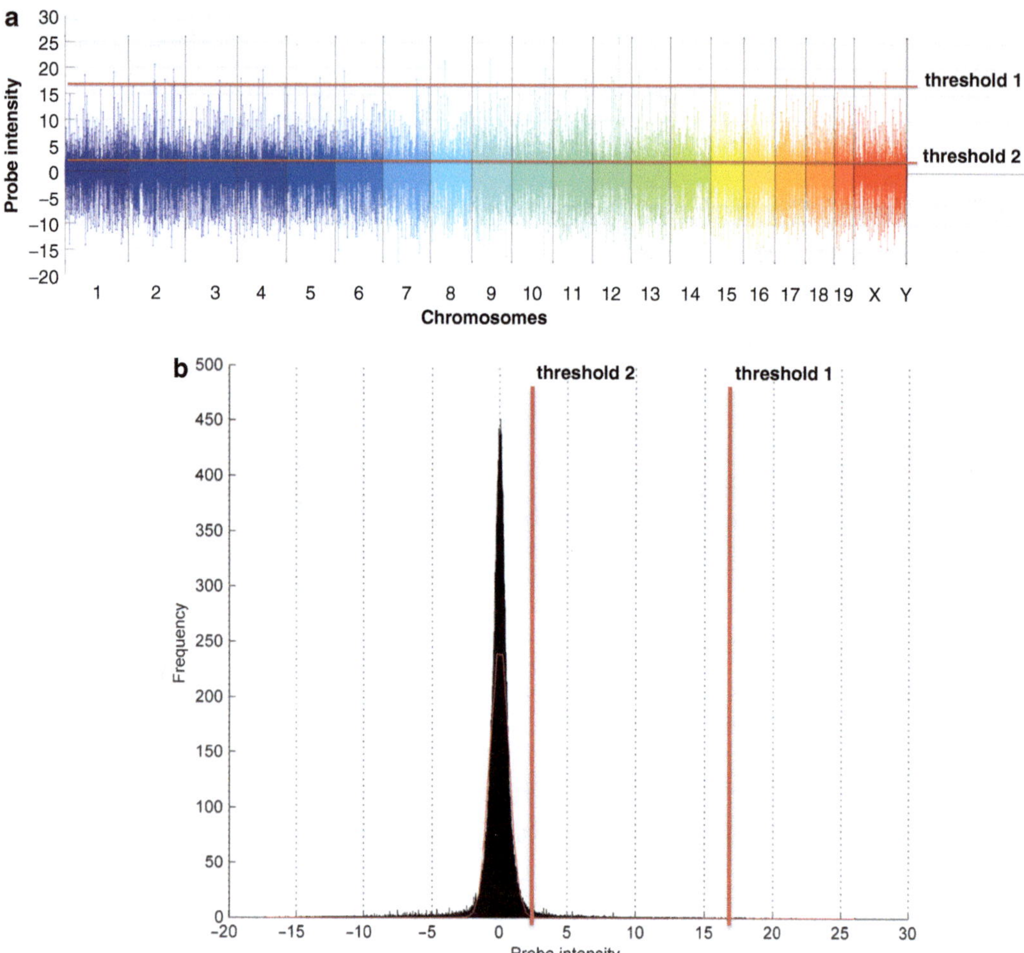

Fig. 1 Graphical representation of probe intensities in a ChIP-on-chip tiling-array experiment. (**a**) Normalized probe intensity of 4.6 million probes among 22 chromosomes. Colors illustrate the different chromosomes whereas the length of a lollipop represents the probe intensity. (**b**) Distribution of the probe-intensity values. The probe-intensity values are normalized against a reference file. Threshold 1 indicates a high threshold cutoff, whereas threshold 2 indicates a low intensity cutoff. HAT uses many different threshold cutoffs to determine significantly enriched ROI

useful for genome-wide studies, the coverage of the genome on the arrays depends on the species that is being studied. As an example, probes can cover the majority of a small genome such as for Arabidopsis [4], whereas probes will cover only contigs in a large genome, such as for human. Thus for larger genomes, as is the case for mouse or humans, the choice of the content depends on the questions one wishes to address using a particular tiling array.

Each tiling-array produces quantitative signal intensity for each probe by the hybridization of labeled DNA. Normalized probe intensities are illustrated by the different peaks in Fig. 1, where the colors indicate the probe signals at different chromosomes.

Although single probe hybridization with high signal intensity suggests strong hybridization, it is not necessarily the result of specific hybridization of labeled DNA (illustrated by the probes above threshold 1 in Fig. 1a, b). Multiple contiguous probes that show increased signal intensity upon hybridization across a particular genomic region are more likely to be the result of true hybridization in a biological experiment. These genomic regions are denoted as a putative region-of-interest (ROI). In order to find such ROI, a low threshold must be employed which may compromise the results by introducing false-positives ROI (Fig. 1a, b, threshold 2). To detect biological relevant ROI, probe-intensity signals should be discriminated from nonspecific signals. A challenge in the analysis of tiling-array data is the detection of true ROI, and to minimize the number of false positives. A straightforward approach is to choose a fixed number of consecutive probes above a certain threshold and indicate it as an ROI. Nevertheless, this definition of ROI may be inadequate because of the required number of consecutive probes and the optimal threshold may be difficult to establish. In addition, the probe resolution varies across the genome, and across different tiling-array platforms.

Multiple methods have been developed to analyze tiling-array data which all serve one goal, i.e., the detection of true ROI and thereby discriminating positive-probe intensity from the background. The developed methods differ in their statistical approaches: methods incorporate the hypergeometric distribution [5], hidden Markov models [6–8], correlation structures [9], heuristics [10], mixture models [11], Bayesian modeling [12, 13], wavelets [14], or by using other methodologies [15–22]. All methods have shown to be useful in filtering large data sets for candidate gene discovery. It is of importance to note that biological experiments are always a necessity to validate particular findings.

Here we discuss the previously developed method, Hypergeometric Analysis of Tiling arrays (HAT) [5], that uses the hypergeometric distribution to assess the probability of a consecutive number of probes in a particular genomic region while controlling multiple testing (Family Wise Error: FWER). Furthermore, HAT uses multiple threshold cutoffs, it does not necessarily require experimental replicates, and can be normalized against reference files. It furthermore employs a single user defined parameter: the significance level alpha. Note that alpha is not used to determine the threshold cutoff using the data distribution (Fig. 1b), instead it computes the probability to observe a specific number of probes for a particular genomic region (window) over multiple threshold cutoffs. Furthermore, specifying parameters such as fragment-size may improve the detection of ROI, whereas parameters for gene-mapping and sequence-of-interest are required for additional analysis (Fig. 2). HAT is generically built and therefore independent of probe-intensity distribution, probe-sets coverage

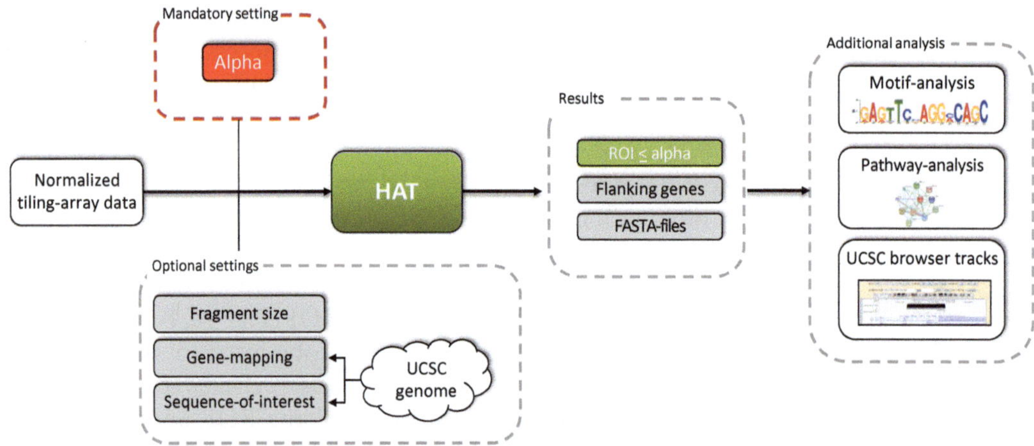

Fig. 2 Schematic overview of tilling-array data analysis. Stepwise illustration of normalized tiling-array data towards the detection of significantly enriched ROI, the flanking genes, sequences files (FASTA), motif analysis, pathway analysis, and the UCSC-browser

and probe-sets resolution across the genome and tiling-array platform. It is successfully applied in multiple types of biological research questions, i.e., the detection of protein–DNA interactions (ChIP-on-chip [5]), identification of genomic locations that are involved in viral integration and potentially harbor tumor suppressor genes (MeDIP-on-chip) [2], the identification of regions enriched for histone modifications such as trimethylation of histone 3 at lysine 4 or lysine 27 (H3K4 me3, H3K27 me3) [5], and for the identification of anthocyanin-specific genes that flank enriched genomic DNA in black rice using 3′-TILLING 135 K *Oryza sativa* microarray [23]. Many detected ROI among these different studies were confirmed by quantitative polymerase chain reaction (qPCR).

Although tiling arrays have been applied successfully for genome-wide applications, high-throughput sequencing of for instance chromatin immunoprecipitated DNA-fragments (ChIP-Seq) show genome-wide associations in higher resolutions and will therefore be superior to chip technology. Even though ChIP-Seq is becoming the standard for genome-wide applications, numerous high-quality tiling-array data sets are publicly available at the gene expression omnibus website (GEO: http://www.ncbi.nlm.nih.gov/geo/). These can be of value to address particular research questions raised by investigators and to which HAT may be very useful. Furthermore, although HAT was initially developed for the analysis of tiling-array data, the application is not limited to the studies discussed in this chapter, but can be applied for the analysis of ChIP-Seq data as well.

Here we stepwise discuss how to apply HAT to analyze tiling-array data. As case examples we used two publicly available ChIP-on-chip data sets. In addition we discuss two types of analyses that frequently follow upon the detection of ROI, namely motif and pathway analyses.

2 Materials

We previously reported the successful usage of HAT on two novel data sets [5]. Here we demonstrate HAT on previously reported STAT4-chromatin immunoprecipitation (ChIP-on-chip) experiments ($n=2$), compared to controls ($n=2$). Secondly, we use HAT to analyze the DNA-binding capacity of a C-terminal mutant *CEBPA* ($n=2$), compared to controls (ER) ($n=2$). Both data sets are available on the gene expression omnibus (GEO), GSE19321 and GSE16845, respectively. Data were generated using the Affymetrix GeneChip Mouse Promoter 1.0 Array. This chip generates 4.6 million perfect match probes over 28,000 mouse promoter regions. Promoter regions cover 6 kb upstream to 2.5 kb downstream of 5′ transcription start sites. Each probe has a size of 25 nucleotides (nt). RAW probe-intensity values are normalized by utilizing Model-based analysis of tiling arrays for ChIP-chip (MAT) [22, 24].

3 Analyzing Tiling-Array Data Sets

In this paragraph we demonstrate the usage of HAT for the identification of significant ROI and define the parameters for ChIP-on-chip experiments. Before starting the peak-detection algorithm (HAT), pre-knowledge about the experimental setup is highly recommended. The experimental protocol requires shearing of the DNA by using a sonication process which results in DNA-fragments of approximately 600 base pairs (bp). Subsequently, chromatin fragments are *immunoprecipitated* using antibodies directed to the protein of interest, known to interact with DNA. The consecutive probes can, therefore, cover up to 600 bp after the hybridization process per fragment. This information can be used in the model for the detection of ROI. Note that significant ROI can be detected that are larger or smaller in width than 600 bp. In addition, we set the significance level on 0.05.

The first ChIP-on-chip data set to which we applied HAT is a study that was previously reported and in which STAT4-mediated transcriptional regulatory networks in Th1 cell development were investigated [1]. STAT4 is a critical component in the development of inflammatory adaptive immune responses. Although STAT4 was subject in various other studies [25, 26], it was claimed that the genetic program, activated by STAT4 that results in an inflammatory cell type, is not well characterized. A ChIP-on-chip experiment was therefore conducted as previously reported [1]. Here, we analyzed both experimental replicates by choosing a fragment-size of 600 nt and α: 0.05, and detected $n=2{,}903$ and $n=3{,}106$ ROI. Moreover, 84 % ($n=2{,}499$) overlapped in both replicates compared to the controls (sized between 215 and

Table 1
Motif enrichment analysis on the detected regions-of-interest in the STAT4 experiment

Transcription factor	Recognized factors	Fold-increase	*P*-value
V$STAT1_01	STAT1, STAT1alpha,STAT1beta	8,0588	5,5308E-29
V$STAT5B_01	STAT5A, STAT5B	4,0922	1,052E-26
V$STAT1_05	STAT1	5,6114	2,1506E-26
V$STAT_01	STAT1, STAT1alpha, STAT1beta,STAT2, STAT3, STAT3-isoform1,STAT4, STAT5A, STAT5B, STAT6	3,5424	5,4465E-22
V$STAT3_01	STAT3, STAT3-isoform1	5,9629	1,3443E-19
V$STAT1STAT1_Q3	CBF3, STAT1:STAT1, ehf	4,0458	1,1127E-18
V$IRF_Q6	IRF-10, IRF-2, IRF-3, IRF-4, IRF-5, IRF-6, IRF-7, IRF-7A, IRF-7B, IRF-7H, IRF-8, IRF4-1, irf1	3,603	1,0817E-11
V$AP1 _Q6_01	AP-1, FOSB, FosB, Fra-1, Fra-2, JunB, JunB:Fra-1, JunB:Fra-2, JunD, JunD:Fra-2, JunD:deltaFosB, c-Fos, c-Jun, c-Jun:FosB, c-Jun:JunD, c-Jun:c-Fos, deltaFosB	2,7031	2,0413E-11
V$STAT5A_01	STAT5A	4,0795	3,8123E-11
V$BACH1_01	Bach1, Bach1t	3,1209	3,3523E-09

The top ten enriched TFBS among the detected binding regions using HAT for the STAT4 study (ChIP-on-chip). A TFBS is called when the position weight matrices (PWM) is enriched at $P \leq 0.001$. Recognized factors: the transcription factors that are recognized by the TFBS. Fold-increase: the frequency that a TFBS is detected among the binding regions compared to the reference set (5,000 randomly chosen genes)

4543 nt, median: 1002 nt). It was previously demonstrated that the analysis method, GenPathway, identified 4,669 genes that were seen in both replicates [1]. This list is subsequently filtered for genes with binding intensity >4 and thereby resulted in 1,540 genes. This indicates that using the unfiltered list, GenPathway detects almost twice the number of ROI when compared to HAT. To investigate the validity of the ROI that were detected by HAT, a motif enrichment analysis was conducted on the 2,499 common ROI by using F-MATCH [27, 28]. We hypothesize that the detected ROI should contain a STAT-binding site. We detected a total of 38 transcription factor binding sites (TFBS) of which the STAT-motifs were highly enriched ($P<0.001$). Moreover, 7 STAT-motifs were detected in the top 10 after ranking the TFBS on significance (Table 1). This suggests high specificity of the detected ROI. Note that the STAT-motif is also highly enriched in the genes detected by GenPathway [1]. Although both methods detected high enrichment for the STAT-motifs, the overlap of genes between both methods was 897 genes. In other words, 1,211 genes were

Table 2
Motif enrichment analysis on the detected regions-of-interest that are exclusively detected using HAT in the STAT4 experiment

Transcription factor	Recognized factors	Fold-increase	*P*-value
V$STAT1_01	STAT1, STAT1alpha, STAT1beta, STAT2, STAT3, STAT3-isoform1, STAT4, STAT5A, STAT5B, STAT6	7,235	4,96E−16
V$STAT3_01	STAT3, STAT3-isoform1	5,862	7,36E−12
V$STAT5B_01	STAT5A, STAT5B	3,6333	2,16E−11
V$STAT1_05	STAT1	4,6577	2,45E−09
V$STAT_01	STAT1, STAT1alpha, STAT1beta	3,4047	4,05E−09
V$GADP_01	GABP	4,4162	1,88E−08
V$SAP1A_01	SAP-1a	4,1057	5,32E−08
V$STAT1STAT1_Q3	CBF3, STAT1:STAT1, ehf	3,5072	5,26E−07
V$ELK1_02	Elk-1, Elk1-isoform1	4,0252	7,68E−07
V$CETS1P54_01	Ets-1, Ets-1 deltaVII, c-Ets-1, c-Ets-1 54, c-Ets-1A, c-Ets-1B	3,9813	8,78E−07

The top ten enriched TFBS among the exclusively detected binding regions of HAT for the STAT4 study (ChIP-on-chip). A TFBS is called when the position weight matrices (PWM) is enriched at $P \leq 0.001$. Recognized factors: the transcription factors that are recognized by the TFBS. Fold-increase: the frequency that a TFBS is detected among the binding regions compared to the reference set (5,000 randomly chosen genes)

solely detected by HAT and not by GenPathway. To assess the validity of these ROI, we conducted a motif analysis for only those 1,211 ROI and detected again high enrichment for the STAT-motifs, i.e., 6 STAT-TFBS are detected in the top ten ranked list (Table 2). We hypothesize that these 1,211 genes may be present in the initial 4,669 genes detected by GenPathway, but are excluded from the list as these did not comply the above mentioned criteria. This is supported by the notion that significantly lower probe-intensity levels are observed ($P < 0.0001$) in the 1,211 ROI compared to the 897 ROI. Note that the probe-intensity levels, of all the detected ROI, are significantly higher compared to the background. Unfortunately, we were not able to analyze the motifs among the exclusively detected genes by GenPathway, as the exact genomic positions of the ROI were not specified. These differences may occur due to alternatively defined gene-mapping procedures (Fig. 3) and the differences in statistical methodologies. In conclusion, we identified another set of genes that were highly enriched for the STAT-motif.

The second ChIP-on-chip data set is used to study the DNA-binding capacity of a variant of CCAAT enhancer binding protein

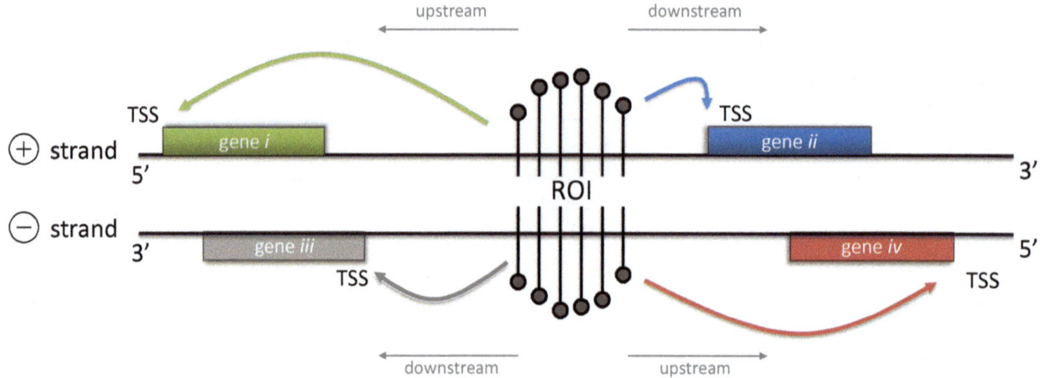

Fig. 3 Mapping of detected regions-of-interest to genes located in close vicinity. A single ROI is illustrated with four neighboring genes: two on the positive strand (upstream and downstream) and two on the negative strand (upstream and downstream). Mapping of ROI to genes is crucial for additional analysis (e.g., pathway analysis). ROI: region-of-interest, TSS: transcriptional start site, 3′: three prime UTR, 5′: five prime UTR

alpha (*CEBPA*) that carries a C-terminal mutation. *CEBPA* is a transcription factor and master regulator of myeloid differentiation [29, 30]. It is frequently mutated in patients with acute myeloid leukemia (AML) (5–14 %) [31]. Abnormalities in *CEBPA* may contribute to a block in differentiation of progenitor cells of granulocytes, which can result in leukemogenesis. Mutations in *CEBPA* are associated with a particular prognosis of patients with AML [31]. In AML patients, two types of *CEBPA* mutations are known to exist: mutations in the N-terminus and the C-terminus. C-terminal mutations are found in the DNA-binding domain. Since the mutant protein can still interact with other proteins that may interact with DNA, we propose that mutant *CEBPA* may indirectly interact with DNA. We wondered to which loci mutant *CEBPA* might interact in an indirect manner. We created a similar C-terminal mutation as found in one particular human AML patient [32], with an insertion of six amino acids in the C-terminal bZIP domain. We used it in the ChIP-on-chip experiment to identify genes that may play a role in leukemogenesis. Promoter array hybridizations were conducted from a myeloid cell line model (32D) that expresses either beta-estradiol inducible C-terminal mutant *CEBPA* (2 clones) or control-ER (2 clones). The question that we wished to address is whether mutated *CEBPA* can bind to the DNA, thereby identifying the associated genes. Using a fragment-size of 600 bp and an alpha of 0.05, we detected in total $n = 89$ and $n = 109$ significant binding regions in the two clones with C-terminal mutant *CEBPA* that was not seen in the controls (Fig. 4). The ROI are sized between 154 and 2,481 nucleotides (median 717 nt) and forty-eight were commonly detected in both clones.

We next searched for binding-motifs among the detected ROI of the C-terminal mutant *CEBPA*. Although it is known that the

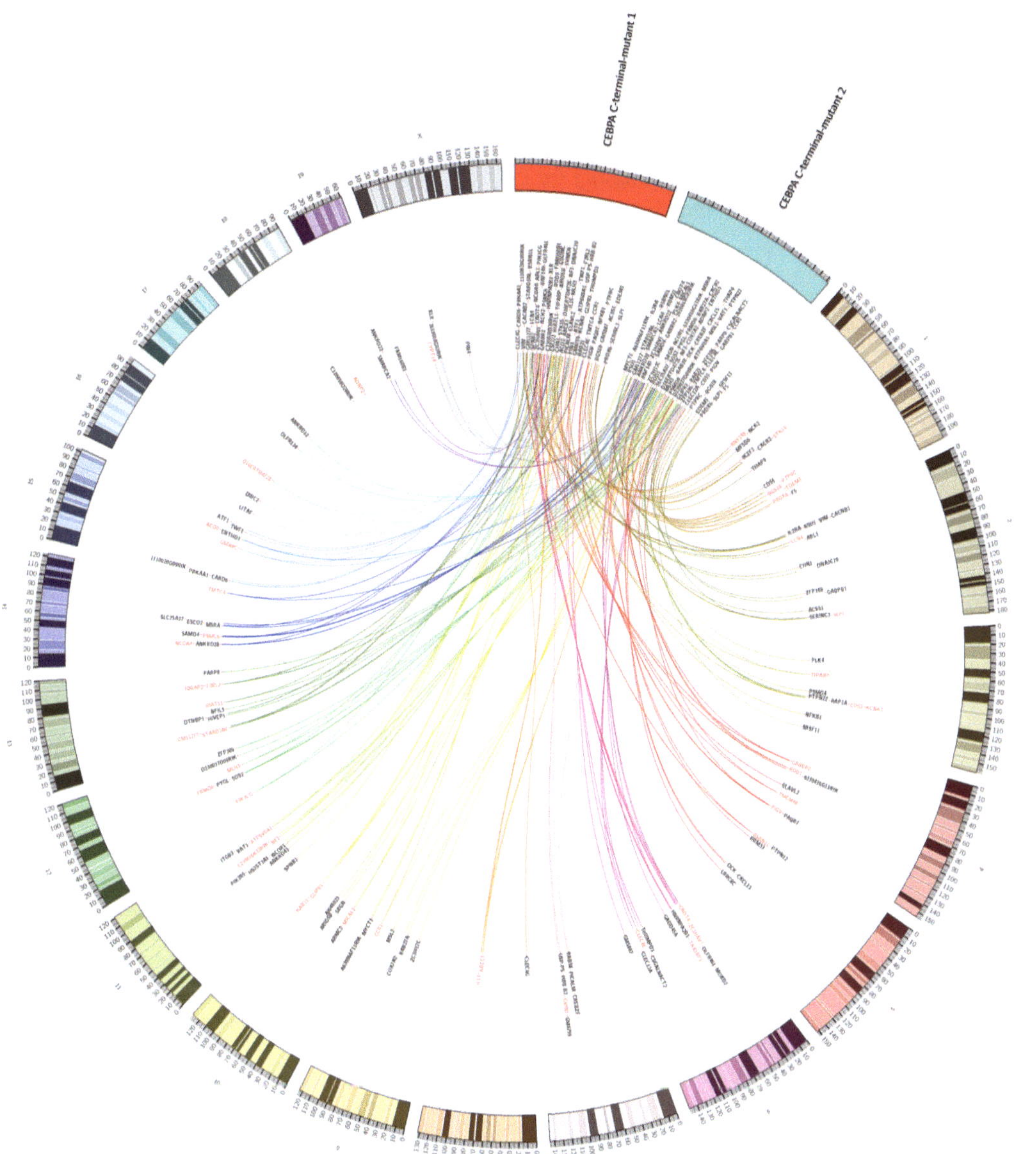

Fig. 4 Graphical representation of the genes that are bound by the C-terminal mutant *CEBPA*. One hundred and forty mapped genes from the detected ROI of the C-terminal mutant *CEBPA* experiments are illustrated. Candidate genes in experiment 1 are indicated by the *red box*, whereas the candidate genes from experiment 2 are indicated by the *blue box*. Forty-six genes (mapped from 48 ROI) that overlap between experiment 1 and 2 are indicated with a *red text-color*. *Line-colors* are colored similar as the chromosomes which are numbered from 1 to 19 and X, and show the relative location of the genes using mouse genome-build 8 (mm8)

C-terminal mutant *CEBPA* lacks binding capacity, we identified three enriched motifs namely, core-binding factor (CBF), ETS, and ESE-1 (P-value < 0.001). Core-binding factors have been shown to fulfil an important role in hematopoiesis [33] and ETS family members, such as ESE-1, fulfil an important role in several

Table 3
Motif enrichment analysis on the 2 kb upstream regions-of-interest of the C-terminal mutant *CEBPA* genes

Transcription factor	Recognized factors	Fold-increase	*P*-value
V$POU3F2_02	POU3F2, POU3F2 (N-Oct-5a), POU3F2 (N-Oct-5b)	1,9479	6,8068E−12
V$CDP_02	CDP, CDP-isoform1, CDP2	2,0723	9,8643E−09
V$FOXP3_Q4	FOXP3	4,753	5,9801E−08
V$OCT1_01	Oct-1, POU2F1, POU2F1a	1,7131	6,2367E−08
V$IPF1_Q6	PDX1, ipf1	1,6051	1,3651E−07
V$CLOX_01	Cutl	1,7556	8,1928E−07
V$SATB1_01	CBF-C	1,5865	8,4804E−07
V$OTX_Q1	Otx1, Otx2	1,9029	9,0873E−07
V$HMGIY_Q3	HMGI-C, HMGIY, HMGIY-isoform1, HMGIY-isoform2	1,546	1,7164E−06
V$FOXO1_01	FOXO1A	1,7142	1,9312E−06
V$DMRT4_01	DMRT4	1,581	2,2881E−06
V$NFAT_Q6	NF-AT, NF-AT1, NF-AT1C, NF-AT2, NF-AT3, NF-AT4, NFAT1, NFAT1-is&$$$;	3,009	2,7411E−06
V$TEF_Q6	TEF-xbb1, Thyrotroph embryonic factor, Thyrotroph embryonic factor&$$$;	1,9549	5,1394E−06
V$SRF_C	SRF, SRF-I, SRF-L, SRF-M, SRF-S	2,1108	5,9055E−06
V$CEBPGAMMA_Q6	C/EBPgamma	1,8284	6,5027E−06

The top 15 enriched TFBS among the 2 kb upstream genes of the C-terminal mutant ROI. A TFBS is called when the position weight matrices (PWM) is enriched at $P \leq 0.001$ and with fold-increase >1.5. Recognized factors: the transcription factors that are recognized by the TFBS. Fold-increase: the frequency that a TFBS is detected among the binding regions compared to the reference set (5,000 randomly chosen genes)

signal transduction pathways [34–36]. As expected, we did not find the consensus binding motif CEBP as we showed previously for wild-type *CEBPA* using the same model system [5]. The detection of these three enriched motifs and the absence of the CEBP motif suggest that DNA-binding by mutant CEBPA had occurred indirectly. We hypothesized that other factors may influence the DNA-binding capacity and therefore analyzed the 2 kb upstream regions, from the transcriptional start site (TSS) of the detected genes (Fig. 4). This resulted in the detection of 71 enriched TFBS with $P \leq 0.001$ and 1.5 times more frequently observed than in the reference set (fold-increase ≥ 1.5). As a reference set we selected 2 kb upstream sequences (starting from the transcription start site) of 5,000 randomly selected genes. The 2 kb upstream sequences are gathered using the UCSC database (http://hgdownload.cse.ucsc.edu). The top 15 TFBS are depicted in Table 3.

3.1 Detected Regions-of-Interest Can Be Mapped to Genes that Are Located in Close Vicinity

Although the goal is to detect ROI by using ChIP-on-chip tiling arrays, it often requires additional analysis, such as pathway analysis, to test a particular hypothesis. This requires the mapping of ROI to genes. Each ROI can, theoretically, be mapped to four genes that are located on: (1) the positive strand and upstream, (2) the positive strand and downstream, (3) the negative strand and upstream, and (4) the negative strand and downstream (Fig. 3). From these four genes, only one gene may be targeted (or two genes in a bi-directional promoter region). For promoter tiling arrays, where only the promoter regions are present on chip, it is straightforward to map the detected ROI to the nearest located TSS of a gene. To prevent incorrect gene-mapping, due to differences in genomic locations of TSS between species and/or genomic-build (hg18, hg19 for human and mm8, mm9 for mus musculus), it is highly recommended to use the same species and genomic-build for both the gene-mapping file as the one used in the normalization process. These gene-mapping files can be downloaded from the UCSC: http://hgdownload.cse.ucsc.edu.

Manually curating each detected ROI to a particular gene is possible using the UCSC-genome-browser track (generated using HAT, Fig. 2) but can be time-consuming. Alternatively, by specifying the species and genome-build in HAT, each ROI can automatically be mapped to the TSS of a gene in closest vicinity. We specified in both ChIP-on-chip experiments "mm8" because the experimental samples were derived from mus musculus and normalized with genomic-build 8. Because both analyzed data sets have been generated using promoter tiling arrays, it allowed the mapping of the ROI to genes in close vicinity. For the STAT study, the 2,499 detected ROI were mapped to 2,108 unique genes. For the *CEBPA* study, it resulted in the detection of 140 unique genes. These are graphically illustrated using a circos-plot [37] (Fig. 4). Such graphical representation indicates the chromosomal location of the genes, and whether genes are commonly detected in the independent experiments using different clones.

3.2 Motif and Pathway Analyses on the Detected Regions-of-Interest and Their Flanking Genes

Analysis on the detected ROI or the genes that are located in close vicinity of the ROI is an important next step for hypothesis testing. Both motif and pathway analyses are therefore useful in tiling-array studies (Fig. 2).

Motif analysis detects specific sequences involved directly in protein–DNA binding interactions, or alternatively whether the promoter regions of the flanking genes include overrepresented sequences of transcription factors. These so-called TFBS may suggest that the protein-of-interest interacts synergistically with other proteins or is involved in the formation of protein complexes. In general, two types of motif analysis exist: by using known TFBS that are derived from published collections (e.g., JASPAR or TRANSFAC databases). These databases should be used when seeking specific factors or structural classes. Secondly, dé-novo motif analysis can be used to analyze similarities among the sequences to produce a description for each pattern it discovers. F-MATCH [27, 28] and MEME [38] are two algorithms which can be used for the detection of known TFBS and/or dé-novo motifs. These methods are online accessible and require FASTA-files as an input, which contain sequences of the ROI (generated by HAT).

Besides motif analysis, it can be useful to analyze the detected genes for enriched pathways. Pathway analysis is the process of identifying interactions and associated annotations [39]. For the detected flanking genes it may provide insight how genes are regulated and which processes, functions, or networks were involved. Both commercial and noncommercial entities provide pathway analysis. A commercial tool is Ingenuity Pathway Analysis (Ingenuity® Systems, http://www.ingenuity.com, IPA 8.8). Networks in IPA are created using literature-based records that are maintained in the Ingenuity Pathway Knowledge Base. It computes a network-score for the overlap of the focus genes with a global molecular network. Alternatively, Gene Set Enrichment Analysis (GSEA) [40] provides both software and a collection of annotated gene sets (MSigDB: Molecular Signature Database) that can be used for the detection of pathways and/or gene sets (noncommercial). Depending on the research question, different gene sets can be used: (1) BioCarta pathways, describing the molecular relationships derived from active research areas, (2) KEGG pathways, describing the molecular interactions and reaction networks, (3) Reactome pathways, manually curated and peer-reviewed pathways, (4) GO biological processes, gene sets describing the biological process ontology, (5) Transcription factor targets (TFT), gene sets contain genes that share a TFBS, and (6) MicroRNA targets, Gene sets that contain genes that share a 3′ UTR microRNA binding motif.

4 Notes

Different Methodologies Come to Different Results, What Is the Correct One to Choose?

All previously described methods have been reported to validate some of the detected ROI as described in Subheading 1. Nevertheless, different statistical methodologies lead to differences in the detected ROI. We hypothesize that various methodologies may results in similar detected ROI which are most likely the genomic regions that contain a contiguous number of probes with high probe-intensity levels (the results of two methods are shown in Subheading 3). In addition, the differences between detected ROI among various methodologies are likely the genomic regions with subtle changes in probe-intensity levels. Note that some developed methodologies are designed for the analysis of one type of tiling-array application. Others may require various parameters to set before starting the analysis, e.g., by defining the ROI using the maximum and/or minimum number of probes in a genomic region, maximum gap size between two probes and threshold. Changing one of the parameters will affect the final results. It is therefore always recommended to perform additional analysis after the detection of ROI to ensure confidence about the gained results. We demonstrated this in Subheading 3, where we detected 1,211 ROI that were exclusively found for HAT. A motif analysis showed significant enrichment for the STAT-consensus binding site. Such findings may help deciding which method to use. It is important to note that in the end laboratory experiments are indispensable to demonstrate the biological significance of particular that ROI, identified by means of tiling-array analysis.

How to Continue If No Significant Regions-of-Interest Are Detected?

The analysis of tiling-array data (Subheading 3) can result in the absence of significantly enriched ROI. This indicates that probe-intensity values, by the hybridization of DNA-fragments on chip, showed no significant differences compared to the background data-file. In case the hybridization process on chip is successfully performed (i.e., DNA-fragments are immunoprecipitated) and the background data-file is correctly provided into the model, it still may result in the absence of significantly enriched ROI. Note that analyzing experimental data-files without the usage or incorrect usage of a background data-file can lead to the absence of significantly enriched ROI or the detection of false-positive ROI. If no significantly enriched ROI are detected, it should be considered that no DNA-binding did take place and therefore no ROI were detected. Alternatively, one could decide to increase the significance level alpha and rerun the analysis. Note that the false-positive rate increases by using alpha >0.05. It is therefore highly recommended to validate the ROI by qPCR. As an example, it is demonstrated that a MeDIP-on-chip experiment resulted in the detection of 15 ROI [2]. These are detected without using a background [2]. Although there was supporting evidence that all 15 ROI may be valid (additional analysis showed that all ROI contained a nearby restriction site), only eight viral integration sites could be validated by directed PCR followed by Sanger sequencing [2]. The remaining seven ROI may therefore be the result of technical variation which may have been prevented by using a background file.

How to Determine the Best Flanking Gene for a Detected Regions-of-Interest?

The usage of tiling-array data does not provide information regarding the strand (positive or negative) or genes affected by the putative promoter. It only indicates the probe-intensity values and their genomic positions. If a particular genomic region is marked as a potential ROI, the responsible immunoprecipitated DNA-fragment is suggested to show binding, e.g., via an immunoprecipitated transcription factor, that could bind to the DNA strand. The ROI is then linked to the gene in close vicinity (Fig. 3). The use of an UCSC-browser track may help manually curating the ROI to a gene. Alternatively, it requires biological experiments to validate whether the binding had an effect on the regulation of a gene. Note that promoter tiling arrays (as described in Subheading 2) only contain probes of which the genomic locations are in the promoter regions of genes and therefore simplifies the gene-mapping procedure.

How to Run HAT with RAW Cell Files?

HAT is build generically to analyze different applications and platforms of tiling-array data (as described in Subheading 1). On the contrary, normalization may differ between different applications and platforms of tiling arrays, e.g., one-color arrays of Affymetrix versus two-color arrays of Nimblegen. Including a normalization step into the model would therefore limit the model to one type of tiling array. RAW cell files need to be normalized based on the type of tiling array [24], and then used as an input into the model (Fig. 2).

How to Prevent "Out-of-Memory" Problems When Analyzing Tiling-Array Data?

When using HAT, it is recommended to use at least 4GB of RAM memory and Windows-64bits version or UNIX-based system. The methodology is tested on tiling-array data containing 4.6 million perfect match probes, and developed in such a way that it is analyzed per chromosome which reduces high memory loads. Nevertheless, when memory problems occur, it is recommended to kill unused running processes when running HAT. In a Windows environment this can be done in the "task manager"; find the "Run" window in the start-menu en type "taskmgr" and then press "Ok" or press the <ENTER>-key.

How to Install HAT in Windows or an UNIX Environment?

The installation of HAT requires an x86-64 Windows or UNIX-based system and 4GB memory or more is highly recommended. Both platforms require the installation of MATLAB or the MATLAB Compiler Runtime (MCR) which is a standalone set of shared libraries that enable full functioning of HAT. The MCR installer and HAT can be downloaded from http://www.erasmusmc.nl/hematologie/.

Windows x86-64 Platform

1. Download HAT and the MCRinstaller setup-file.
2. Run the setup-file and check the option: "MCR-package for win64."
3. After the installation, the MCR directory (<mcr_root>\<version>) will automatically be included in the PATH. Alternatively the environment variable can be set using the command-prompt (**3.1**) or without the command-prompt (**3.2**).
   3.1. Open a command-prompt and issue the command:

   set PATH=<mcr_root>\<version>\runtime\win64;%PATH%

   3.2. For Windows, add the PATH environment variable as follows.

- Select the My Computer icon on your desktop or in the configuration window.
- Right click the icon and select Properties from the menu.
- Select the Advanced tab.
- Click Environment Variables.
- Click "Edit" for variable "PATH"
- Add the "<mcr_root>\<version>\runtime\win64" to the "PATH" (delimited by semicolons (;)).
- Run HAT by executing "HAT.exe".

UNIX x86-64 Platform

1. Download the HAT and MCRinstaller file.
2. Install the MATLAB Compiler Runtime with the following command.

```
./MCRInstaller.bin -is:extract
./MCRInstaller.bin -console
```

For a noninteractive and non-GUI installation.

```
./MCRInstaller.bin -P bean421.installLocation="desiredInstallPath" -silent
```

3. After the installation, add the MCR directory (<mcr_root>/<version>) to the environment variable.

```
setenv LD_LIBRARY_PATH
<mcr_root>/<version>/runtime/glnxa64:
<mcr_root>/<version>/bin/glnxa64:
<mcr_root>/<version>/sys/os/glnxa64:
<mcr_root>/<version>/sys/java/jre/glnxa64/jre/lib/amd64/native_threads:
<mcr_root>/<version>/sys/java/jre/glnxa64/jre/lib/amd64/server:
<mcr_root>/<version>/sys/java/jre/glnxa64/jre/lib/amd64:
setenv XAPPLRESDIR<mcr_root>/<version>/X11/app-defaults
```

4. Run HAT by issuing the command: "./HAT".

Acknowledgment

We wish to thank Claudia Erpelinck-Verschueren for technical assistance in the preparation of the *CEBPA* C-terminal mutant samples.

References

1. Good SR, Thieu VT, Mathur AN, Yu Q, Stritesky GL, Yeh N et al (2009) Temporal induction pattern of STAT4 target genes defines potential for Th1 lineage-specific programming. J Immunol 183:3839–3847
2. Beekman R, Valkhof M, Erkeland SJ, Taskesen E, Rockova V, Peeters JK et al (2011) Retroviral integration mutagenesis in mice and comparative analysis in human AML identify reduced PTP4A3 expression as a prognostic indicator. PLoS One 6:e26537
3. Crawford GE, Davis S, Scacheri PC, Renaud G, Halawi MJ, Erdos MR et al (2006) DNase-chip: a high-resolution method to identify DNase I hypersensitive sites using tiled microarrays. Nat Methods 3:503–509
4. Yamada K, Lim J, Dale JM, Chen H, Shinn P, Palm CJ et al (2003) Empirical analysis of transcriptional activity in the Arabidopsis genome. Science 302:842–846
5. Taskesen E, Beekman R, de Ridder J, Wouters BJ, Peeters JK, Touw IP et al (2010) HAT: hypergeometric analysis of tiling-arrays with application to promoter-GeneChip data. BMC Bioinformatics 11:275
6. Munch K, Gardner PP, Arctander P, Krogh A (2006) A hidden Markov model approach for determining expression from genomic tiling micro arrays. BMC Bioinformatics 7:239
7. Yu WH, Hovik H, Chen T (2010) A hidden Markov support vector machine framework incorporating profile geometry learning for identifying microbial RNA in tiling array data. Bioinformatics 26:1423–1430
8. Knott SR, Viggiani CJ, Aparicio OM, Tavare S (2009) Strategies for analyzing highly enriched IP-chip datasets. BMC Bioinformatics 10:305
9. Kuan PF, Chun H, Keles S (2008) CMARRT: a tool for the analysis of ChIP-chip data from tiling arrays by incorporating the correlation structure. Pac Symp Biocomput 515–526
10. Zhang Y (2008) Poisson approximation for significance in genome-wide ChIP-chip tiling arrays. Bioinformatics 24:2825–2831
11. Wu H, Ji H (2012) JAMIE: a software tool for jointly analyzing multiple ChIP-chip experiments. Methods Mol Biol 802:363–375
12. Mo Q, Liang F (2010) A hidden Ising model for ChIP-chip data analysis. Bioinformatics 26: 777–783
13. Mo Q, Liang F (2010) Bayesian modeling of ChIP-chip data through a high-order Ising model. Biometrics 66:1284–1294
14. Karpikov A, Rozowsky J, Gerstein M (2011) Tiling array data analysis: a multiscale approach using wavelets. BMC Bioinformatics 12:57
15. Otto C, Reiche K, Hackermuller J (2012) Detection of differentially expressed segments in tiling array data. Bioinformatics 28:1471–1479
16. Lan X, Bonneville R, Apostolos J, Wu W, Jin VX (2011) W-ChIPeaks: a comprehensive web application tool for processing ChIP-chip and ChIP-seq data. Bioinformatics 27:428–430
17. Kechris KJ, Biehs B, Kornberg TB (2010) Generalizing moving averages for tiling arrays using combined p-value statistics. Stat Appl Genet Mol Biol 9:Article29
18. Zacher B, Kuan PF, Tresch A (2010) Starr: simple tiling ARRay analysis of Affymetrix ChIP-chip data. BMC Bioinformatics 11:194
19. Droit A, Cheung C, Gottardo R (2010) rMAT–an R/Bioconductor package for analyzing ChIP-chip experiments. Bioinformatics 26:678–679
20. Judy JT, Ji H (2009) TileProbe: modeling tiling array probe effects using publicly available data. Bioinformatics 25:2369–2375
21. Cesaroni M, Cittaro D, Brozzi A, Pelicci PG, Luzi L (2008) CARPET: a web-based package for the analysis of ChIP-chip and expression tiling data. Bioinformatics 24:2918–2920
22. Johnson WE, Li W, Meyer CA, Gottardo R, Carroll JS, Brown M, Liu XS (2006) Model-based analysis of tiling-arrays for ChIP-chip. Proc Natl Acad Sci USA 103:12457–12462
23. Kim CK, Kikuchi S, Hahn JH, Park SC, Kim YH, Lee BW (2010) Computational identification of anthocyanin-specific transcription factors using a rice microarray and maximum boundary range algorithm. Evol Bioinform Online 6:133–141
24. Bolstad BM, Irizarry RA, Astrand M, Speed TP (2003) A comparison of normalization methods for high density oligonucleotide array data based on variance and bias. Bioinformatics 19:185–193
25. Lieberman LA, Banica M, Reiner SL, Hunter CA (2004) STAT1 plays a critical role in the regulation of antimicrobial effector mechanisms, but not in the development of Th1-type responses during toxoplasmosis. J Immunol 172:457–463
26. Nguyen KB, Watford WT, Salomon R, Hofmann SR, Pien GC, Morinobu A (2002) Critical role for STAT4 activation by type 1

interferons in the interferon-gamma response to viral infection. Science 297:2063–2066

27. Kel AE, Gossling E, Reuter I, Cheremushkin E, Kel-Margoulis OV, Wingender E (2003) MATCH: a tool for searching transcription factor binding sites in DNA sequences. Nucleic Acids Res 31:3576–3579
28. Kel A, Voss N, Jauregui R, Kel-Margoulis O, Wingender E (2006) Beyond microarrays: find key transcription factors controlling signal transduction pathways. BMC Bioinformatics 7(Suppl 2):S13
29. Rosenbauer F, Tenen DG (2007) Transcription factors in myeloid development: balancing differentiation with transformation. Nat Rev Immunol 7:105–117
30. Friedman AD (2007) Transcriptional control of granulocyte and monocyte development. Oncogene 26:6816–6828
31. Taskesen E, Bullinger L, Corbacioglu A, Sanders MA, Erpelinck CA, Wouters BJ et al (2011) Prognostic impact, concurrent genetic mutations, and gene expression features of AML with CEBPA mutations in a cohort of 1182 cytogenetically normal AML patients: further evidence for CEBPA double mutant AML as a distinctive disease entity. Blood 117:2469–2475
32. van Waalwijk B, van Doorn-Khosrovani S, Erpelinck C, Meijer J, van Oosterhoud S, van Putten WL, Valk PJ et al (2003) Biallelic mutations in the CEBPA gene and low CEBPA expression levels as prognostic markers in intermediate-risk AML. Hematol J 4:31–40
33. de Bruijn MF, Speck NA (2004) Core-binding factors in hematopoiesis and immune function. Oncogene 23:4238–4248
34. Rabault B, Ghysdael J (1994) Calcium-induced phosphorylation of ETS1 inhibits its specific DNA binding activity. J Biol Chem 269: 28143–28151
35. Tenen DG, Hromas R, Licht JD, Zhang DE (1997) Transcription factors, normal myeloid development, and leukemia. Blood 90:489–519
36. Grall FT, Prall WC, Wei W, Gu X, Cho JY, Choy BK et al (2005) The Ets transcription factor ESE-1 mediates induction of the COX-2 gene by LPS in monocytes. FEBS J 272: 1676–1687
37. Krzywinski M, Schein J, Birol I, Connors J, Gascoyne R, Horsman D et al (2009) Circos: an information aesthetic for comparative genomics. Genome Res 19:1639–1645
38. Bailey TL, Elkan C (1994) Fitting a mixture model by expectation maximization to discover motifs in biopolymers. Proc Int Conf Intell Syst Mol Biol 2:28–36
39. Viswanathan GA, Seto J, Patil S, Nudelman G, Sealfon SC (2008) Getting started in biological pathway construction and analysis. PLoS Comput Biol 4:e16
40. Subramanian A, Tamayo P, Mootha VK, Mukherjee S, Ebert BL, Gillette MA et al (2005) Gene set enrichment analysis: a knowledge-based approach for interpreting genome-wide expression profiles. Proc Natl Acad Sci USA 102:15545–15550

Chapter 10

Inference of Alternative Splicing from Tiling Array Data

Johannes Eichner

Abstract

Alternative splicing (AS) is an important mechanism implicated in eukaryotic gene expression, whereby exon segments of precursor-mRNA transcripts are joined together in different arrangements corresponding to diverse isoforms of mature mRNA. Accumulating evidence suggests that in many instances this process is specifically regulated and contributes to the structural and functional diversification of tissues and cell types. Furthermore, several studies support the view that environmental stresses dramatically impact on AS and reported the presence of novel transcript isoforms in response to biotic or abiotic stresses. Since specific regulation of AS in plants is a largely unexplored field of research, large-scale approaches aimed at monitoring AS on a genome-wide level are of increasing importance to gain insights into tissue-specific splicing regulation and to study the effects of changed environmental conditions on pre-mRNA splicing.

Here, we describe the concepts of a traditional statistical approach, and a more recently developed machine learning-based method for AS detection from tiling arrays. The here presented approaches were employed for the detection and profiling of AS events in the model plant *A. thaliana*, and applied to a large dataset comprising transcriptomic expression data from 11 tissues and 13 stress conditions.

Key words Alternative splicing, Exon skips, Intron retentions, *Arabidopsis thaliana*, Tiling arrays, Machine learning, Support Vector Machines

1 Introduction

1.1 Alternative Splicing

Alternative splicing (AS) is an important posttranscriptional mechanism, which diversifies gene functions and thereby increases the complexity of the transcriptome and the proteome. The splicing of pre-mRNAs is catalyzed by the spliceosome, a large macromolecular ribonucleoprotein complex, consisting of four small nuclear ribonuclein particles (snRNPs) and about 300 non-snRNP proteins [1]. The spliceosome accurately recognizes splice sites, excises introns, and joins the exons together. In higher eukaryotes protein-coding genes are often composed of multiple exons, which can be joined in different arrangements to form mature mRNA transcripts. The respective proteins, encoded by distinct mRNA isoforms from the same gene, mostly have similar functions. However, they can differ in characteristics, such as binding

Tin-Lap Lee and Alfred Chun Shui Luk (eds.), *Tiling Arrays: Methods and Protocols*, Methods in Molecular Biology, vol. 1067, DOI 10.1007/978-1-62703-607-8_10,

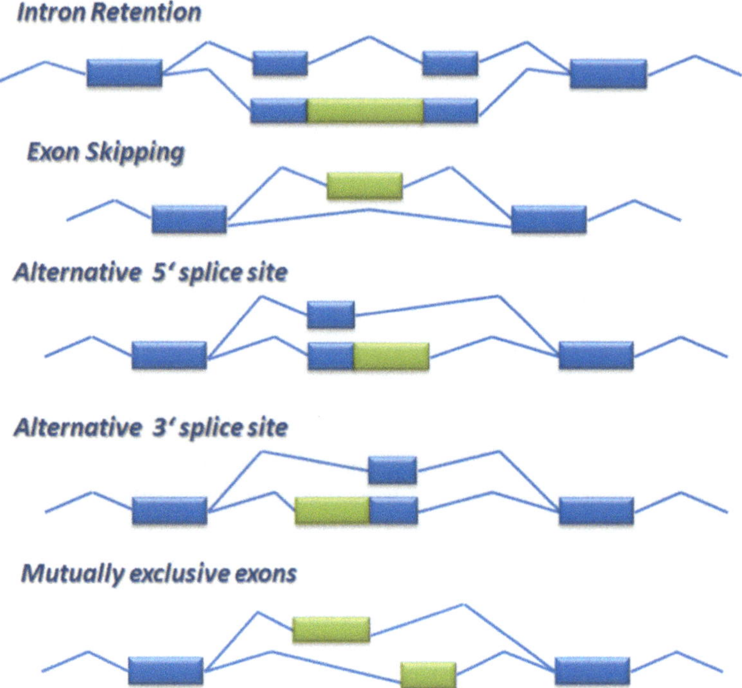

Fig. 1 Modes of alternative splicing. The most common modes of alternative splicing are shown by means of splicing graphs. Pre-mRNA segments, which are included in mature mRNA transcripts are represented by *blue boxes* and segments, which are excised before nuclear export are indicated by *blue lines*. Isoform-specific segments, which may vary across transcripts of the same gene, are shown in *green boxes*

specificity, enzymatic activity, stability, and subcellular location [2]. In contrast to constitutive splicing, where all exons of a gene are joined together in a single fixed composition, we define alternative splicing as the mechanism which regulates the generation of distinct mature mRNA transcripts from the same gene by differential use of splice sites. Note that this definition of AS excludes other types of transcript isoform variation, such as alternative transcription start sites or polyadenylation sites. The most common types of AS events are illustrated in Fig. 1. Several studies of AS in different organisms reported that AS events may undergo differential regulation between tissues, i.e., the ratios of alternative transcript isoforms were observed to vary across tissues [3–5], suggesting that tissue-specific differential splicing played a major role in the evolution of specialized cell and tissue types [5, 6]. The regulation of AS is essentially based on interactions between *cis*-acting sequences in the pre-mRNA, i.e., splicing enhancers and silencers, and a variety of *trans*-acting splicing factors, which may exhibit different levels and activities, depending on tissue types and environmental conditions [7]. Biochemical studies of protein–RNA interactions showed that splicing factors, such as serine/arginine-rich proteins (SR proteins), bind to regulatory elements of pre-mRNA transcripts and thereby have an impact on splice site selection [8]. Remarkable effort has been directed at the identification of short, loosely conserved RNA subsequences which correspond to splicing factor binding sites, and the genomic locations and motives of several *cis*-acting elements have been determined in mammals [5]. In animals these elements have been categorized into four classes, depending on their location and effect on pre-mRNA splicing: exonic splicing enhancers (ESE), exonic splicing silencers (ESS), intronic splicing enhancers (ISE), and intronic splicing silencers (ISS). In contrast to animals, only few studies reported the discovery of *cis*-regulatory elements involved in splicing regulation in plants [1]. As computational analysis of plant genes were mostly limited to ESEs [9], the determination of regulatory pre-mRNA elements remains an interesting field for further research. As the number of genes is not correlated with the cellular complexity of eukaryotes, alternative splicing is assumed to account for much of the increased complexity of higher organisms [6, 10]. Furthermore, AS is known to be frequently coupled with nonsense-mediated decay (NMD), and a substantial fraction of alternative splicing events result in transcripts containing a premature termination codon (PTC), which potentially triggers NMD degradation. In *A. thaliana* about 40 % of the AS events result in mRNA transcripts harboring a PTC [11]. Diverse studies of AS in human reported that most transcripts with a PTC located more than fifty nucleotides upstream of an exon–exon junction are potential targets of NMD [12, 13]. This mechanism, which causes unproductive splicing in the sense that the affected transcript is degraded after one round of translation, has been shown to regulate gene expression in eukaryotes on the post-transcriptional level [13]. Hence, AS is assumed to provide another layer of gene regulation, which accounts for the fine-tuning of gene expression [1, 6]. Moreover, AS plays a central role in the regulation of development and cellular differentiation, and misregulation of AS is known to impact on physiological processes and has been associated with life threatening human diseases [14].

1.2 Splicing-Sensitive Microarray Platforms

Tiling arrays with splicing level resolution provide genome-wide measurements of gene expression which facilitate the development of accurate AS detection methods and thereby contribute to the elucidation of tissue-specific and stress-induced splicing regulation. Many of the limitations inherent in sequence-based approaches have been overcome by microarrays, which enable unbiased whole-genome studies of AS [3–5, 10]. An overview of the most common microarray types employed for profiling AS can be found in Fig. 2. Microarrays designed for AS detection enable the identification and quantification of transcript isoforms on a genome-wide level. Depending on their design this is even possible for transcripts not

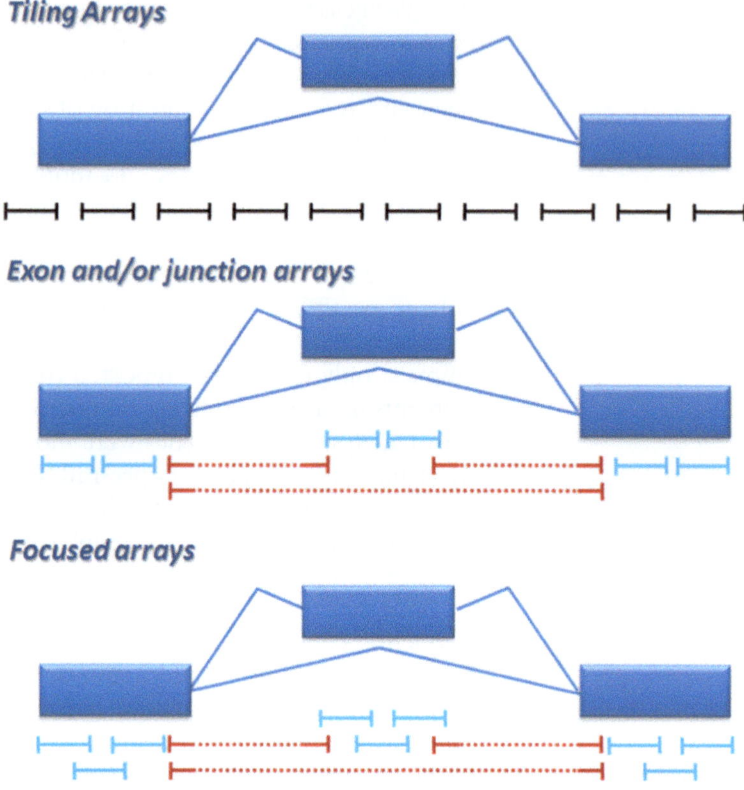

Fig. 2 Microarray platforms for AS detection. The figure illustrates the design of AS microarray probes complementary to a gene composed by two constitutive and one alternatively spliced exon. *Black lines* indicate tiling probes, which are uniformly spaced and spread over the whole non-repetitive portion of the genome. *Blue lines* correspond to exon body probes and exon junction probes are drawn in *red*. Note that the focused and exon junction array platform provide more evidence than tiling arrays for the transcript isoform in which the internal exon is spliced out, because of the gain of signal at the exon junction probe between the initial and terminal exon

known a priori. Due to their highly parallel nature, microarrays are ideal tools for profiling gene expression and splicing under a variety of environmental conditions. Since the resolution of commonly used gene-centered microarrays is generally not high enough to accurately detect AS events, special microarray platforms were introduced to discover transcript isoform variation. Most microarrays employed for monitoring AS events can be assigned to one of the following categories: tiling arrays, exon or junction arrays, and focused arrays [15]. Generally all of these microarray platforms are well suited for the identification and quantification of mRNA splice forms. However, each platform has its advantages and limitations, making them more or less appropriate for certain organisms or detection tasks. Exon or junction arrays as well as tiling arrays enable the discovery of new splice forms and typically provide robust quantitative information [15]. The essential difference of the two platforms is that tiling arrays are designed independently of gene annotations, whereas exon arrays require a priori knowledge about exon boundaries and the design of exon junction probes typically even incorporates knowledge about alternative transcripts. In contrast to exon or junction arrays which are restricted to the detection of splicing events affecting exons, such as alternative 5′ and 3′ ends or complete exon skips, tiling arrays facilitate the *ab initio* detection of any known form of AS or more general transcript isoform variation. Thus, tiling arrays can also be employed to monitor intron retention events [16], which are known to be the predominant type of AS in the model plant *A. thaliana* [11, 17]. Since it was shown that exon skips are the most prevalent AS event in human [18], and due to the large size of mammalian genomes consisting mostly of intergenic and intronic sequence, studies of AS in mammals typically employed exon junction arrays [3] or microarrays comprising both exon and exon junction probes [4, 5]. As the design of tiling probes is more highly constrained than for exon probes, robust expression measurements require slightly larger probesets, as probe-specific effects and crosshybridization are more likely affecting these measurements [19]. Furthermore, the extremely large number of probes makes computational analyses more challenging [15]. However, exon junction arrays suffer from similar problems, as junction probes must conform design specifications and are therefore likely to be affected by crosshybridization and often exhibit suboptimal hybridization properties. In contrast to tiling arrays and exon arrays which are aimed at monitoring the whole transcriptome, focused arrays are specially designed for the quantification of known splice forms of specific groups of genes. Due to the focused design, high probe coverage generally allows for accurate quantitative analyses, and data analysis is relatively simple, because of the small number of comprised probes compared to tiling arrays. However, the design of focused arrays heavily relies on the complete annotation of well-characterized genes, making it an ideal tool for diagnostic purposes, while discovery of new transcript isoforms is beyond the scope of its application.

1.3 Microarray-Based Studies of Alternative Splicing

Johnson et al. published a pioneering, custom microarray-based study of AS in human, which was aimed at elucidating the biological function of exon skipping events by large-scale profiling of AS [3]. The authors monitored AS across 52 tissues and cell lines using exon junction arrays and discovered numerous AS events that have not yet been detected by sequence-based approaches. They showed that ESTs are biased against AS detection in low-abundance transcripts, substantiated by an increase in EST-based AS detection rate with high expression levels. Furthermore they demonstrated that their exon junction arrays are sensitive to AS events occurring in genes whose expression level is well below average and that the AS detection rate is largely independent of the expression level. Pan et al. employed focused microarrays comprising both exon body and junction probes [4], realizing the improvements suggested by Johnson et al. [3]. For the array design they restricted their study to sequence-verified exon skips and mutually exclusive exons, and profiled AS across various tissues and cell types. The authors developed a new analysis tool GenASAP [10], which quantitatively determines inclusion and exclusion levels of alternatively spliced exons, using machine learning and Bayesian network approaches. Their sophisticated model accounts for expression-dependent noise and outliers, i.e., probes showing aberrant signal levels. In principle the tool is also applicable to quantify other types of AS events, such as intron retentions, provided that focused array platforms are used. Sugnet et al. also used custom-designed focused oligonucleotide microarrays, comprising both exon and exon junction probes to analyze differential splicing across 22 adult mouse tissues [5]. Their microarrays contain isoform-specific probesets and one additional common probeset per gene measuring overall gene expression. The common probeset comprises probes complementary to constitutive exons and was used to detect which genes exhibit expression above background, based on a p-value computed from an empirical cumulative distribution function. To detect differentially spliced exons they used two transcript-specific probesets per event, which comprise both exon body and junction probes for inclusion, and distinct exon junction probes for skipping of particular exons. Using the Kruskal-Wallis test, they determined alternative transcript isoforms whose intensity distributions at junction probes are significantly different across tissues, and inferred alternative splicing if both RNA forms are differentially expressed. The authors focused on exon skips whose splicing pattern, i.e., inclusion or skipping, is different in brain or muscle, relative to other tissues, using a regression-based bootstrapping method. The identification of sequence motifs of splicing factor binding sites which are involved in brain and muscle-specific regulation of splicing, respectively, was achieved by a comparison of differentially included and constitutive exons. For this purpose the conservation of intron sequences adjacent to exons, which are included or skipped in brain or muscle tissues, was compared to the flanking sequences of constitutive exons. Ner-Gaon and Fluhr employed whole-genome tiling arrays to monitor intron retention events in the model plant *A. thaliana* [16]. They performed a genome-wide survey of intron retention events using an approach based on statistical tests, and verified their results by two different validation strategies, namely RT-PCR experiments and a comparison against the TAIR6 annotation. The RT-PCR validation yielded an estimated precision of 86 %, whereas the verification of the predictions against TAIR6 resulted in an overlap of 18 %. Moreover, they discovered characteristic features of retained introns, such as higher GC content, smaller size, and enrichment in untranslated regions. These studies demonstrate that tiling arrays are an accurate tool for large-scale profiling of AS on a genome-wide level. Furthermore, AS microarrays provide a discovery platform for the identification of previously unknown AS events and typically allow to precisely characterize the experimental conditions under which the AS events occur. However, microarray-based approaches are also known to have several drawbacks. Due to the limited sensitivity of microarrays, the detection of low abundant minor transcript isoforms is exceedingly difficult and hardly feasible [16]. Like ESTs, tiling arrays do not allow to determine whether two AS events in the same gene occur in the same or in different transcript isoforms unless they are differentially spliced and measured. Moreover, AS detection by microarrays mostly requires differential expression, i.e., many of the presently available AS detection tools, such as the Affymetrix tool MIDAS (http://www.affymetrix.com/support/technical/whitepapers/exon_alt_transcript_analysis_whitepaper.pdf), cannot detect AS, if the isoforms are present at the same ratios in all analyzed samples.

2 Materials

We analyzed two tiling array datasets, generated from plant material of *A. thaliana* (Col-0 referenced strain). The first dataset contains hybridization data from 11 tissues and developmental stages of *A. thaliana* [20]. The second one contains stress response data from five abiotic stress conditions, each measured at two points of time (1 and 12 h after treatment) [21]. All samples from the stress response dataset were derived from *A. thaliana* plant seedlings. For all plant tissues and stress conditions, samples from three biological replicates were prepared.

2.1 Arabidopsis Tissue Atlas Dataset

The following description of the plant material and tissue sample preparation was taken from Laubinger et al. [20]:

1. Wild-type Col-0 and clv3-7 plants [22] were grown on soil or on solid MS medium under continuous light at 23 °C. Table 1 contains a description of the analyzed hybridization samples.
2. Tissue samples were frozen in liquid nitrogen and total RNA was isolated using the RNeasy Plant Mini Kit (Qiagen, Hilden, Germany).
3. RNA integrity was determined on a Bioanalyzer with the RNA 6,000 Series II Nano kit (Agilent, Santa Clara, CA, USA).

Table 1
Description of samples in *Arabidopsis* tissue atlas dataset [20]

Sample	Genotype	Description
1	WT (Col-0)	Roots, 7 days
2	WT (Col-0)	Seedlings, aerial parts, 7 days
3	WT (Col-0)	Expanding leaves, 10 days
4	WT (Col-0)	Senescing leaves, 35 days
5	WT (Col-0)	Stem, second internode
6	WT (Col-0)	Vegetative shoot meristem, 7 days
7	WT (Col-0)	Inflorescence shoot meristem, 21 days
8	WT (Col-0)	Whole inflorescences to floral stage 9, 21 days
9	*clv*3-7 (Col-0)	Whole inflorescences (*clv*3-7), 21 days
10	WT (Col-0)	Flowers, stage 15, >21 days
11	WT (Col-0)	Fruits, carpels stage 15, >21 days

2.2 Arabidopsis Stress Response Dataset

The following description of the stress response sample preparation was taken from Zeller et al. [21].

1. Arabidopsis seeds (wild-type Col-0 strain) were plated on half-strength MS medium with 1 % sucrose.
2. The Arabidopsis seeds were kept for 3 days at 4 °C on the plates.
3. The plates were transferred in continuous light at 21 °C.
4. After 10 days a control sample was taken (time point zero control).
5. Plants were transferred to liquid MS medium with 1 % sucrose (Mock control).
6. MS medium was supplemented 200 mM NaCl (salt stress), 300 mM mannitol (osmotic stress), and 100 μM ABA (plant hormone).
7. Plants were kept at 8 ± 1 °C (cold stress) or 30 ± 1 °C (heat stress) in precooled and pre-warmed liquid MS medium.
8. Samples were taken after 1 and 12 h and frozen in liquid nitrogen (Table 2).

2.3 Array Design

The expression data was measured by the Affymetrix GeneChip Arabidopsis Tiling 1.0R Array, which comprises more than 3.2 million perfect match probes, and as many mismatch probes tiled throughout the whole non-repetitive portion of the *A. thaliana* genome. The central positions of the adjacent 25-mer probes are spaced 35 base pairs on average, leaving a gap of approximately ten base pairs between the probes.

Table 2
Description of samples in *Arabidopsis* stress response dataset [21]

Sample	Time point	Description
1	t_0 (0h)	Negative control
2	t_1 (1h)	Negative control
3	t_1 (1h)	Salt stress (200 mM NaCl)
4	t_1 (1h)	Osmotic stress (300 mM mannitol)
5	t_1 (1h)	Plant hormone (100 μM ABA)
6	t_1 (1h)	Cold stress (8 ± 1 °C)
7	t_1 (1h)	Heat stress (30 ± 1 °C)
8	t_2 (12h)	Negative control
9	t_2 (12h)	Salt stress (200 mM NaCl)
10	t_2 (12h)	Osmotic stress (300 mM mannitol)
11	t_2 (12h)	Plant hormone (100 μM ABA)
12	t_2 (12h)	Cold stress (8 ± 1 °C)
13	t_2 (12h)	Heat stress (30 ± 1 °C)

3 Methods

In this chapter we provide a detailed description of the protocol used in the microarray experiments and the bioinformatics pipeline used for the preprocessing, annotation, and computational analysis of the data (Fig. 3). Here, the main focus of the analysis was laid on the inference of diverse AS events from the tiling array data (Fig. 4).

3.1 Probe Preparation and Array Hybridization

The preparation of the samples and the hybridization to the tiling arrays were performed as described previously [20, 21]. For the sake of completeness, we again provide a step-by-step description of the whole procedure.

1. For synthesis of probes (targets) for tiling arrays, 1 μg of total RNA was used as template for generation of cRNA using the MessageAmp II-Biotin Enhanced Kit (Ambion, Austin, TX, USA).
2. We followed the manufacturer's instructions with one exception; biotinylated NTPs were replaced by unmodified NTPs (stock solution 25 mmol/l each).
3. Seven micrograms of unmodified cRNA (for tiling arrays) was converted into dsDNA (GeneChip® WT Double-Stranded

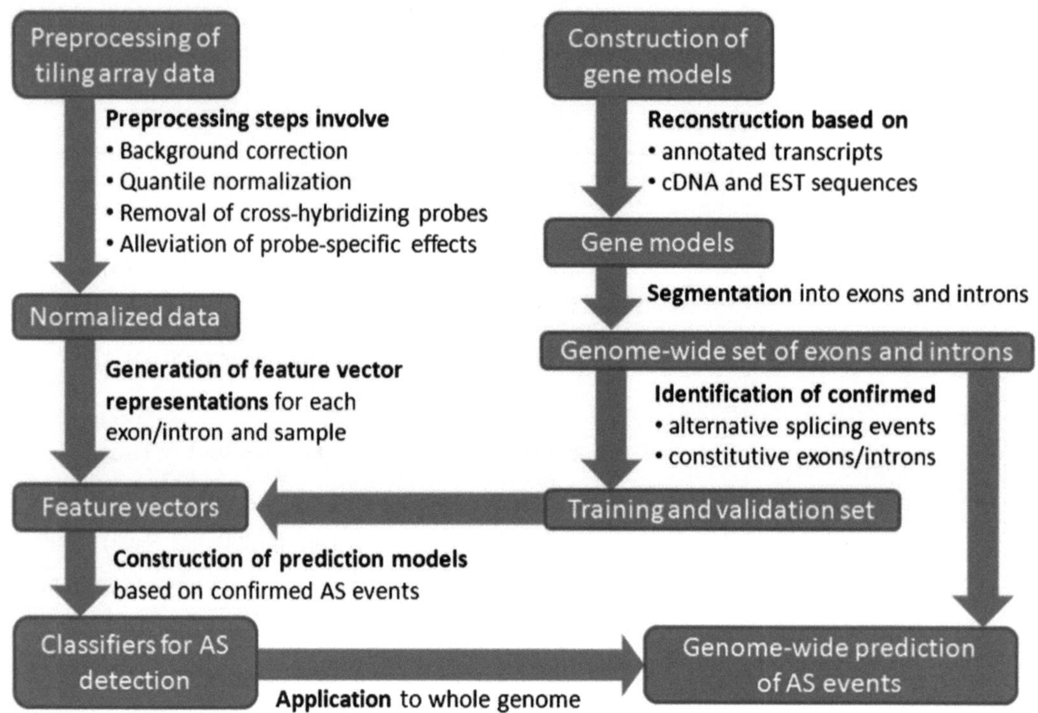

Fig. 3 Bioinformatics pipeline for the inference of AS from tiling arrays. The figure shows a schematic overview of the methodology proposed for the detection of alternative splicing events from whole-genome tiling arrays. First different preprocessing steps are necessary to correct for diverse experimental artifacts present in the raw data. In order to map the raw data to the genome, we propose the generation of gene models based on annotated transcripts and cDNA/EST sequences. These gene models are then fragmented into exons and introns for which AS shall be investigated. A set of sequence-confirmed alternatively and constitutively spliced exons and introns, respectively, is compiled based on the transcripts reconstructed for each gene. An appropriate feature representation capturing the intensities of the complementary probes has to be generated for each exon/intron and sample. Classifiers for AS detection can then be constructed and evaluated based on the training and validation set, respectively. Finally, the genome-wide inference of AS events can be performed based on the generated prediction models

cDNA Synthesis Kit; Affymetrix Inc.) and dsDNA was purified using the MinElute Reaction Cleanup Kit (Qiagen).

4. A total of 7.5 μg dsDNA were fragmented and labeled using the GeneChip® WT Double-Stranded DNA Terminal Labeling Kit (Affymetrix, Inc.).
5. Targets were hybridized to Arabidopsis Tiling 1.0R arrays for 14 h at 42 °C, washed (Fluidics Station 450, wash protocol FS450_0001) and scanned using a GeneChip® Scanner 3,000 7 G.
6. For comparison of polyA(+) and polyA(±), rRNA was depleted from 10 μg total RNA using RiboMinus™ Yeast Transcriptome Isolation Kit (Invitrogen) and an *Arabidopsis*-specific RiboMinus™ LNA oligonucleotide mix kindly provided by Invitrogen, Carlsbad, CA, USA.

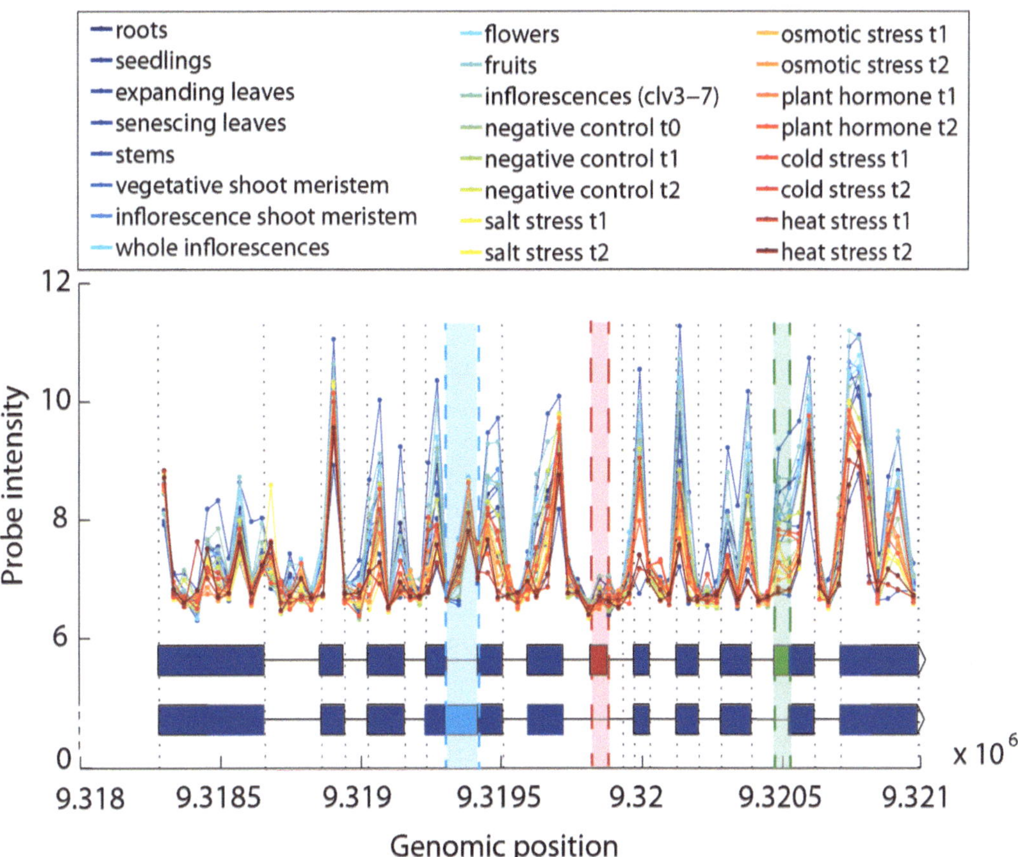

Fig. 4 AS detection from tiling array data. The figure depicts an example gene with two transcripts along with the expression profiles measured in diverse samples using tiling arrays. Exons are depicted as *blue boxes* and introns as *black horizontal lines*. The curves correspond to the probe intensities obtained for different hybridization samples (here: different tissues and stress conditions). Diverse types of alternative splicing events and their typically observed profiles are highlighted (*blue*: intron retention, *red*: exon skip, *green*: alternative 3′ splice site)

7. rRNA depleted RNA was precipitated and resuspended in 12 μl water, from which 11 μl were used for reverse transcription using MessageAmp II-Biotin Enhanced Kit (Ambion) with an oligo-dT-T7 primer (MessageAmp II-Biotin Enhanced Kit) or a random-T7 primer (included in the GeneChip® WT Amplified Double-Stranded cDNA Synthesis Kit; Affymetrix, Inc.).

3.2 Preprocessing of Tiling Array Data

In this subsection, we describe the preprocessing steps which were performed to correct for experimental artifacts, crosshybridization, and probe sequence-specific effects (*see* **Note 1**).

1. *Background correction*. We corrected for the uneven hybridization background of individual microarrays using a technique proposed by Borevitz et al. [23]. At first we determined a background image by calculating the mean $\log_2$-transformed intensity of unmasked features in a quadratic window around each spot.

Subsequently the background image, i.e., the matrix containing the computed means for each spot, was subtracted from the original expression matrix. The resulting spatially corrected hybridization values were then rescaled by adding a constant in log-space to adjust the medians of background corrected intensities to the median of the original raw intensity distribution.

2. *Normalization between arrays.* To correct for the inherent variation of the laboratory experiments, which causes differing intensity distributions between arrays, the distributions of probe intensities were mapped to the empirical mean intensity distribution of all arrays by quantile normalization [24].
3. *Exclusion of ambiguous probes.* In order to avoid the crosshybridization of the 25-mer tiling array probes to multiple transcripts from different genomic locations, we excluded probes if their sequences occurred more than once in the genome. For this purpose we used a method proposed by Clark et al. [25], which detects 25-mers, which occur multiple times in the genome, (1) with identical sequences, (2) with identity subsequences at the central 21 nucleotides, and (3) with a single mismatch.
4. *Sequence normalization.* Probe sequence-specific effects were corrected by a transcript normalization technique proposed by Zeller et al. [26]. The method is aimed at reducing sequence biases by modeling the difference of the observed probe intensities y_i to the ideal, constant exonic or intronic signal level μ_i by a function $f(x_i, y_i) \approx y_i - \mu_i$. The function f depends on the measured intensity and sequence properties x_i, namely position and location of subsequences up to length three, sequence entropy, GC content, and scores based on potential secondary structures. To estimate the ideal exonic signal level μ_i of a transcript, probes were mapped to annotated genes derived from TAIR [27], and μ_i was set to the median intensity of the exonic probes hybridizing to the respective transcript. Ideal intronic signal levels were determined analogously. The probe intensities y_i were discretized into 20 quantiles, and the deviation from the ideal intensity $f(x_i, y_i) = f_q(x_i)$ was estimated separately for each quantile $q \in \{1,\ldots,20\}$ by means of Ridge Regression.

3.3 Definition of Intron Retentions and Exon Skips

Intron retentions (IR) and exon skips (ES) were defined based on the splicing graph representation of genes (*see* Fig. 1) as done by Eichner et al. [28].

1. *Intron retentions* were defined as introns which are spliced out from at least one alternative transcript isoform of a gene, while at least one isoform is known where an exon stretches across the same region.

2. *Exon skips* were defined as exons which are present in at least one isoform, while in at least one alternative isoform an intron stretches across the same genomic region. Additionally, we required that the two alternative isoforms share at least one upstream and one downstream exon, in order to distinguish exon skips from other types of transcript isoform variation, such as alternative transcriptional start sites or polyadenylation sites.

3.4 Inference of AS from Sequence-Based Gene Models

The gene models we used for our computational analyses of AS incorporated TAIR annotated transcripts as well as EST/cDNA sequence information [28].

1. We obtained cDNA and full-length cDNA sequences from the RIKEN Arabidopsis Genome Encyclopedia (RARGE) [29].
2. EST sequences were collected from the dbEST database [30].
3. At first we built a cDNA/EST-based gene structure by using BLAT [31] to align EST/cDNA sequences to the genome of *A. thaliana*. If EST/cDNA sequences matched to different genomic locations, we only accepted the best hit.
4. From the alignments we computed confirmation counts for each intron and exon which correspond to the number of EST/cDNA sequences confirming the two adjacent splice sites.
5. We merged consistent alignments which shared at least one complete exon or intron. For this purpose, alignments were clustered using a bottom-up merging algorithm which iteratively joined two clusters containing sequences that match to the same genomic location.
6. Another set of *Arabidopsis* gene models were parsed from a gff3-file containing the current version of the TAIR annotation [27].
7. The annotation-based gene models were combined with the EST/cDNA-based gene structure by merging overlapping transcripts located on the same strand.
8. Then we built splicing graphs for each gene in which exons correspond to nodes with genomic coordinates and introns to joining edges between exons.
9. In order to measure the reliability of the exons and introns contained in splicing graphs, we computed confirmation counts based on the number of ESTs and cDNAs providing sequence confirmation for the respective adjacent splice sites. To account for the higher quality and the fact that full-length cDNAs provide much more reliable evidence than ESTs, we counted them twice.
10. AS events were inferred from the gene models based on the definitions of IR and ES events given in Subheading 3.3.

Table 3
Datasets used for training and evaluation of the SVM-based AS detection method [28]

Dataset	Positive examples	Negative examples
Constitutive exon/ intron splicing set	71928 annotated constitutively spliced exons	47952 annotated constitutively spliced introns
Alternative intron splicing set	762 IR events with EST/cDNA confirmation for both isoforms	13132 EST/cDNA-confirmed constitutively spliced introns
Alternative exon splicing set	173 ES events with EST/cDNA confirmation for both isoforms	14492 EST/cDNA-confirmed constitutively spliced exons

3.5 Generation of the Validation Set

The validation sets for assessing the prediction accuracy of the implemented AS detection methods were compiled based on sequence-confirmed AS events as previously described by Eichner et al. [28].

1. The IR and ES events detected from splicing graph analysis of the generated gene models were used as confirmed positive examples for our validation set (*see* Table 3).
2. To ensure sufficient quality of the validation set, we included only exons and introns for which a confirmation count ≥2 was computed, i.e., each differentially used splice site had to be supported by either 2 ESTs or 1 cDNA sequence (*see* **Note 2**).
3. For the negative examples, i.e., constitutively spliced exons and introns, we required a confirmation count ≥ 5.
4. As the class distribution of the validation set should ideally resemble the natural frequency of AS in *A. thaliana*, the size of the *constitutive exon/intron splicing set* (*see* Table 3) was preassigned, based on a ratio of alternatively to constitutively spliced genes expected for the whole genome. To this end, we set the proportion of alternatively spliced genes to 30 %, using an overestimation of the proportion estimated by Wang and Brendel [11], who reported that 22 % of the genes in the *A. thaliana* genome are subject to AS. We believe that this ratio is an underestimation of the true ratio [1], as the number of EST and cDNA sequences used for this analysis only amounts to 10 % of the sequences used in a related study in human [32]. This view is also shared by Simpson et al., whose results indicate that one-third of the genes in *A. thaliana* undergo AS [33].
5. As our study was limited to IR and ES events, we determined the expected proportion of these events among all types of AS events, based on the frequencies of different types of sequence-confirmed AS events, obtained from the ASIP database (http://www.plantgdb.org/ASIP/Download/). This proportion was determined as $r_{IE} = 70.9$ %.

6. Then we calculated the proportion of genes which are subject to IR or ES events among all genes as $r = r_{IE} \cdot r_{AS} = 21.3$ %.
7. As our method was not evaluated on a set of genes, but on a set of introns and exons, respectively, we had to determine the class distribution on the exon/intron level from the ratio r which reflects the class distribution on the gene level. For this purpose, we computed the number of IR $i_a = 996$ and ES event $e_a = 259$, which occurred in $g_a = 964$ alternatively spliced genes, based on our splicing graph gene models.
8. On the basis of the estimated number of alternatively spliced genes $g_a = (g_a + g_c) \cdot r = 964$, we calculated the number of single-isoform genes $g_c = 3{,}568$.
9. According to an estimate by Reddy the average gene of *A. thaliana* contains 5 exons and 4 introns. Hence, on the simplifying assumption that there is at the most one AS event per gene, we estimated the number of constitutively spliced exons $e_c = 5 \cdot g_c + 4 \cdot g_a = 21{,}696$ and the number of constitutively spliced introns $i_c = 4 \cdot g_c + 3 \cdot g_a = 17{,}164$.
10. Finally, we obtained a class distribution of i_a: $i_c = 1$: 17 for the *alternative intron splicing set* and e_a: $e_c = 1$: 84 for the *alternative exon splicing set* (*see* Table 3).
11. As our gene models did not contain a sufficient number of constitutively spliced introns and exons with the required sequence confirmation (i.e., confirmation count ≥ 5), the exons/introns were sampled with replacement from a basic population, such that the validation sets could be adapted to our estimates of the natural class distributions.

3.6 ANOVA-Based Approach for the Detection of AS

Traditional approaches for array-based AS detection typically employ statistical methods, which detect IR and ES events based on deviations from their expected intensity distribution. Here, we describe a naïve statistical approach which employs an ANOVA test to detect intron retentions based on differences in the expression levels observed between pairs of samples. The general assumption behind this method is that the expression level of alternatively spliced introns is expected to differ between samples, whereas constitutively spliced introns are expected to show similar expression levels across all samples.

1. For each sample t and exon/intron s covered by the probes $p_1, \ldots, p_n$, we computed an intensity vector representation $I_t(s) = (I_{t1}(p_1), \ldots, I_{t3}(p_n))$, where $I_{tr}(p_i)$ denotes the intensity of probe i in the rth biological replicate of sample t.
2. Then we calculated the median intensity $m_t(g(s)) = median(p_{c(1)}, \ldots, p_{c(k)})$ *of all probes* $p_{c(1)}, \ldots, p_{c(k)}$ located in constitutively spliced exons of the gene $g(s)$ that contains the intron s as an estimate for overall gene expression.

3. In order to compensate for varying overall gene expression levels between samples, we computed splicing index vectors by dividing each component of $I_t(s)$ by the estimated overall gene expression level $m_t(g(s))$.
4. The computed splicing index vectors were compared using an ANOVA test and by imposing a cutoff (e.g., $p(s) < 0.05$) on the resulting *p*-values $p(s)$, we detected introns which are differentially spliced across samples.

3.7 SVM-Based Approach for the Detection of AS

While there is only a limited amount of known AS events which can be used as label information for a machine learning-based AS detection method, the number of reliably annotated constitutive exons and introns is very large. Thus, we formulated the problem of identifying IR and ES events as a supervised learning task and designed a two-stage classification procedure. First, we trained SVMs to distinguish between constitutive exons and introns in single samples. In a second step, the predictions made for single samples were integrated across samples to detect IR and ES events. An overview of the implemented workflow is shown in Fig. 5.

3.7.1 Expression-Dependent Classification

As gene expression can differ by several orders of magnitude between genes and experimental conditions, we propose to train classifiers specialized to certain intervals of global gene expression levels.

1. The training data of the SVMs applied in Stage 1 (see *constitutive exon/intron splicing set* in Table 3) was split into ten disjoint bins, depending on gene expression. The limits of these

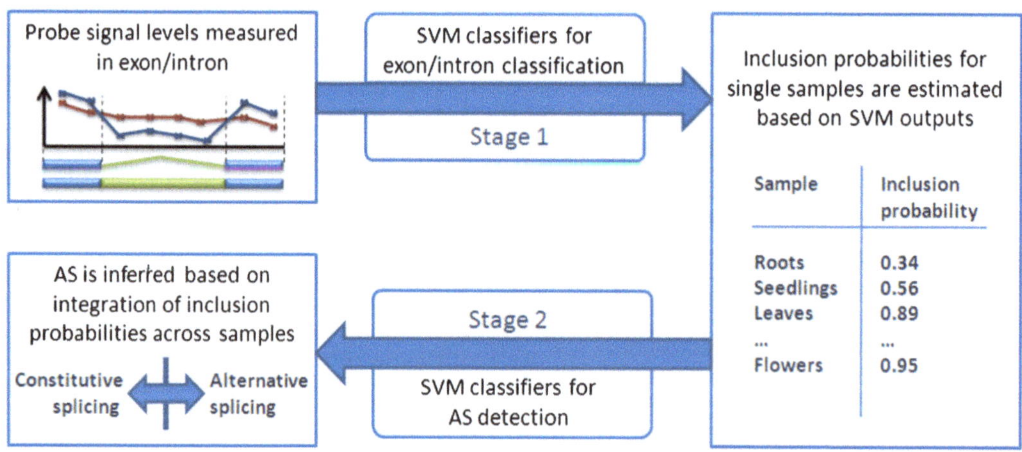

Fig. 5 Workflow of SVM-based AS detection method. In the first stage of the algorithm, SVMs are employed to discriminate exons from introns based on diverse features derived from their hybridization profiles and relative positions in the transcript. For each hybridization sample (i.e., experimental condition) the probability that the given exon/intron is included in mature mRNA transcripts is estimated by transformation of the predicted SVM scores. In the second stage, another set of classifiers is used to infer differential intron and exon splicing, respectively, based on integration of the sample-specific inclusion probabilities which resulted from Stage 1 [28]

expression bins were defined as the percentiles P_{10}, P_{20},..., P_{90} calculated from the distribution of the median exonic probe intensities measured under each condition for each gene.

2. For each *expression bin*, a separate SVM classifier was trained for exon/intron classification at a well-defined range of gene expression (*see* **Note 3**). The details of classifier training are described in following subsections.

3.7.2 Extraction of Features

We computed three different types of features capturing the signal levels and positions of the tiling probes interrogating the expression levels of an exon/intron in a particular sample.

1. *Absolute intensity features* were computed from the absolute expression levels observed for individual exons and introns. For this purpose, we calculated a feature vector $F_{\text{abs}}(s, t)$ for each exon/intron s and each sample t by calculating the percentiles P_{20}, P_{40}, P_{50}, P_{60}, P_{80} of the corresponding intensity vector $I_t(s) = (I_{t1}(p_1), \ldots, I_{t3}(p_n))$, where p_i denotes the ith probe complementary to exon/intron s and $I_{tr}(p_i)$ is its signal level measured in the rth replicate.
2. *Relative intensity* features were extracted from the tiling array data to capture the relative expression level of an exon/intron, compared to the probe signal level distribution observed on a genome-wide level. To this end, we computed the percentiles $L = (P_{20}, P_{40}, P_{50}, P_{60}, P_{80})$ from the global genic probe distribution obtained from the union of all exonic and intronic probe signal levels in all samples combined. Given the vector of probe signal levels $I(s, t) = (I_{t1}(p_1),\ldots,I_{t3}(p_n))$ of a segment s in a sample t, which is interrogated by the probes $p_1,\ldots,p_n$, we computed the components of $F_{\text{rel}}(s, t)$ by linear interpolation between the global percentiles $L_1,\ldots, L_N$. First, we initialized $F_{\text{rel}}(s, t) = (0, 0, 0, 0, 0)$, where each component corresponds to one of the $N = 5$ global percentiles. Then, each value in the intensity vector $I(s, t)$, was assigned to an interval limited by either the first, the last, or two neighboring global percentiles. If the intensity value was lower than the first global percentile $L_1 = P_{20}$, the first component of the feature vector $F_{\text{rel}}(s, t)$ was increased by 1. Analogously, the last component of the feature vector was incremented by one if the intensity value was higher than the last global percentile $L_N = P_{80}$. If the intensity value was assigned to an interval, which is limited by two neighboring global percentiles L_k and L_{k+1}, the kth and $k+1$th components of the feature vector were increased by α and $1 - \alpha$, respectively, where α was computed as defined in equation (3). A formula for the computation of the ith component $F_{\text{rel}}(s, t)_i$ of the feature vector $F_{\text{rel}}(s, t)$ is given in equation (1).

$$F_{\text{rel}}(s,t)_i = \sum_{j=1}^{n} \sum_{r=1}^{3} f\left(I_{tr}\left(p_j\right), i\right) \tag{1}$$

$$f(I,i)=\begin{cases} 1 & \text{if } (I \le L_1 \wedge i=1) \vee (I > L_N \wedge i=N) & [1] \\ \alpha & \text{if } L_k \le I < L_{k+1} \wedge i=k & [2] \\ 1-\alpha & \text{if } L_k \le I < L_{k+1} \wedge i=k+1 & [3] \\ 0 & \text{otherwise} & [4] \end{cases} \tag{2}$$

$$\alpha = \frac{L_{k+1} - I}{L_{k+1} - L_k} \tag{3}$$

3. *Positional features* were incorporated in SVM training to compensate for a bias in the distribution of probe intensities along transcripts, which was observed by Zeller et al. [26]. Specifically, the authors found that the intensities measured at 3′ ends of transcripts are generally higher and decrease towards the 5′ end. A plausible explanation for this phenomenon is a premature termination of transcript amplification, which proceeds from the poly(A) tail of the transcript, resulting in incomplete transcripts. Intending to account for this bias, we provided the SVM classifiers with positional features $F_{\text{pos}}(s)$ encoding the distance of segments to the 3′ ends of the corresponding transcripts. Given the positions of the probes $p_1,\ldots, p_n$ complementary to an exon/intron s, we computed the distances of the probes from the 3′ end of the corresponding spliced mRNA transcript. For an exonic probe p_i this distance equates to the sum of the lengths of the succeeding exons plus the distance of the probe p_i to the next exon/exon junction located downstream of the probe. For intronic probes p_j we added the distance of the p_j to the 3′ donor site of the corresponding intron to the sum of the lengths of the succeeding exons. Next, the positional features $F_{\text{pos}}(s)$ were calculated for each exon/intron s by linear interpolation between arbitrarily defined distance limits. For this purpose, the same algorithm as for the generation of the relative intensity features $F_{\text{rel}}(s, t)$ was applied. As a result we obtained a feature vector $F_{\text{pos}}(s)$ with $N=10$ components corresponding to 10 linearly spaced position limits $L=(100, 300, 500, \ldots, 1{,}900)$.
4. The final feature vector representation $F(s, t)=(F_{\text{rel}}(s, t), F_{\text{abs}}(s, t), F_{\text{pos}}(s))$ for each exon/intron s and sample t was obtained by concatenating the three feature types described above.

3.7.3 Training of SVM Classifiers for Exon/Intron Classification

SVM classifiers were trained separately for each expression bin and a fivefold cross-validation was performed on the *constitutive exon/intron splicing set* (*see* Table 3), in order to optimize the model parameters and to evaluate the classification performance. For this purpose, we used an efficient CPLEX-based implementation (http://www.ilog.com/products/cplex/) of linear SVMs, which solves the underlying convex optimization problem in primal space.

1. The available input and the corresponding labels for each SVM were subdivided into five disjoint sets, of which 60 % were used for training and 20 % for model selection and testing, respectively.
2. During cross-validation, the data subsets were permuted iteratively, such that each instance was part of the test set in one of the five iterations. In each iteration the optimal value of the soft margin parameter C was determined on the validation set based on a grid search ($C \in \{0.001, 0.01, 0.1, 1, 10, 100, 1{,}000\}$) and the prediction accuracy was estimated on the test set, based on ROC scores.
3. The ROC scores obtained for the five disjoint test sets, which resulted from different splits of the data, were subsequently averaged to produce a single assessment of the performance.

3.7.4 Estimation of Inclusion Probabilities for Exons and Introns

In order to gain confidences which are comparable between SVMs, trained on different datasets with different hyperparameters C, we converted the computed SVM scores to inclusion probabilities. The inclusion probability $p_{\text{in}}(s, t)$ for each segment/sample pair (s, t) was determined by transforming the corresponding SVM score into a probabilistic confidence by means of piecewise linear functions.

1. The inclusion probability $p_{\text{in}}(s, t)$ was determined by estimating the conditional likelihood $P(y=1|f_{\text{svm}}(s, t))$ that segment s is included in mature mRNA transcripts expressed in sample t, given the predicted SVM score $f_{\text{svm}}(s, t)$.
2. The SVM scores were discretized into $N=50$ quantiles, which were in turn used as supporting points ϕ_i with $i \in \{1,\ldots,N\}$. For each point ϕ_i we computed the probability $\hat{\pi}_i$ of being a true positive as $\hat{\pi}_i = \frac{n_i^{\text{TP}}}{n_i}$, where n_i denotes the number of instances with SVM scores $f_{\text{svm}}(s, t) \in [\phi_{i-1}, \phi_i]$, and n_i^{TP} is the number of instances with SVM scores in the same interval, which are labeled as exons.
3. To ensure that the piecewise linear functions are monotonically increasing we used a method proposed by Sonnenburg et al. [34]. To this end, the posterior probabilities $P(y=1|f_{\text{svm}}(s, t))$ were computed for every SVM score $f_{\text{svm}}(s, t)$ by linear interpolation between the modified empirical probabilities $\pi_1, \ldots, \pi_N$.

$$p_{\text{in}}(s,t) = P(y=1 \mid f_{\text{svm}}(s,t)) = \begin{cases} \pi_1 & \text{if } f_{\text{svm}}(s,t) < \phi_1 \\ h(\phi_i, \phi_{i+1}) & \text{if } \phi_i \le (s,t) < \phi_{i+1} \\ \pi_N & \text{if } f_{\text{svm}}(s,t) \ge \phi_N \end{cases} \quad (4)$$

$$h(\phi_i, \phi_{i+1}) = \frac{\pi_{i+1}(f_{\text{svm}}(s,t) - \phi_i) + \pi_i(\phi_{i+1} - f_{\text{svm}}(s,t))}{\phi_{i+1} - \phi_i} \quad (5)$$

3.7.5 Training of SVM Classifiers for AS Detection

In the second stage of our supervised AS detection method, the sample-specific inclusion probabilities were incorporated as features by another SVM classifier which was employed for the detection of IR and ES events. For variants of the method which can be used for detecting differential AS between two or multiple experimental conditions the reader is referred to **Note 4**.

1. We restricted our analyses to the 50 % genes with highest expression levels, as the inference of AS requires sufficiently high differences between intronic and exonic probe signal levels. To this end, the median signal level of the exonic probes, pooled across all samples, was used as an estimate for overall gene expression.
2. We trained the Stage 2 SVM classifiers on the single-sample inclusion probabilities estimated in Stage 1, complemented by additional expression features.
3. For each exon/intron an *inclusion feature vector* was generated, which contained the sample-specific inclusion probabilities sorted in descending order, such that the first and last component of the vector corresponded to the sample with maximal and minimal probability, respectively.
4. In the next step we appended *expression feature vectors* of equal length, which incorporated the median signal level measured for the surrounding exons.
5. In order to enable SVM classifiers to learn dependencies between exon/intron inclusion probabilities and expression levels, the *expression feature vector* was permuted such that the order of samples was consistent with the *inclusion feature vector*.
6. We trained two SVM classifiers: one for discriminating retained from constitutively spliced introns and another one for distinguishing ES events from constitutive exon splicing events. Both classifiers were trained and evaluated on the *alternative exon* and *intron splicing set* (*see* Table 3), respectively, using a fivefold cross-validation (*see* **Note 3**).
7. The continuous SVM prediction scores were converted to binary values by selecting the class discrimination cutoff corresponding to maximum precision.

4 Notes

1. As the probe design of tiling arrays is more highly constrained than for ordinary gene-centered arrays, robust expression measurements require slightly larger probesets, as probe-specific effects and crosshybridization are more likely affecting these measurements [19]. Furthermore, additional normalization steps

(e.g., sequence normalization) and filter steps (e.g., filtering of probes with multiple hits in the genome) are required to alleviate probe-specific effects and to avoid crosshybridization, respectively.

2. Even though SVMs are robust against label noise to some extent, the correct labeling of the training instances is crucial for obtaining accurate classifiers. Thus, one should ensure a high quality of the training data.
3. We recommend to scale each feature, such that the mean $\mu = 0$ and the standard deviation $\sigma = 1$ across all instances. If scaled features are used in conjunction with SVM classifiers, numerical instabilities can be avoided.
4. Here, we describe different all-sample scores, which can be employed for the integration of single-sample inclusion probabilities in the second stage of the classification procedure. If differential splicing between multiple experimental conditions (e.g., different tissues) shall be detected, we propose to use the all-sample score defined in equation (6), where $p(s, t)$ denotes the inclusion probability of exon/intron s for stress treatment t. An all-sample score, which is more robust towards outlier samples can be obtained by using the second largest and second smallest value, instead of the maximum and minimum in equation (6). If differential splicing between pairs of conditions is of interest (e.g., stress-treated samples vs. untreated controls) we propose the all-sample score, defined in equation (7). These scores can either replace or complement the SVM-based integration of single-sample scores for AS prediction. In the latter case, an SVM classifier is used to identify AS events, and these are in turn filtered for events showing differential AS using one of the all-sample scores described above.

$$S_{\text{Multi}}(s) = \max_{t \in T}(p_{\text{in}}(s,t)) - \min_{t \in T}(p_{\text{in}}(s,t)) \tag{6}$$

$$S_{\text{Pair}}(s) = \max_{t,c \in T}\left|p_{\text{in}}(s,t) - p_{\text{in}}(s,c)\right| \tag{7}$$

References

1. Reddy ASN (2007) Alternative splicing of pre-messenger RNAs in plants in the genomic era. Annu Rev Plant Biol 58:267–294
2. Stamm S (2002) Signals and their transduction pathways regulating alternative splicing: a new dimension of the human genome. Hum Mol Genet 11:2409–2416
3. Johnson JM, Castle J, Garrett-Engele P, Kan Z, Loerch PM, Armour CD et al (2003) Genome-wide survey of human alternative pre-mRNA splicing with exon junction microarrays. Science 302:2141–2144
4. Pan Q, Shai O, Misquitta C, Zhang W, Saltzman AL, Mohammad N et al (2004) Revealing global regulatory features of mammalian alternative splicing using a quantitative microarray platform. Mol Cell 16:929–941
5. Sugnet CW, Srinivasan K, Clark TA, O'Brien G, Cline MS, Wang H et al (2006) Unusual intron conservation near tissue-regulated exons found by splicing microarrays. PLoS Comput Biol 2:e4
6. Blencowe BJ (2006) Alternative splicing: new insights from global analyses. Cell 126:37–47

7. Simpson CG, Lewandowska D, Fuller J, Maronova M, Kalyna M, Davidson D et al (2008) Alternative splicing in plants. Biochem Soc Trans 36:508–510
8. Black DL (2003) Mechanisms of alternative pre-messenger RNA splicing. Annu Rev Biochem 72:291–336
9. Pertea M, Mount SM, Salzberg SL (2007) A computational survey of candidate exonic splicing enhancer motifs in the model plant Arabidopsis thaliana. BMC Bioinformatics 8:159
10. Shai O, Morris QD, Blencowe BJ, Frey BJ (2006) Inferring global levels of alternative splicing isoforms using a generative model of microarray data. Bioinformatics 22:606–613
11. Wang BB, Brendel V (2006) Genomewide comparative analysis of alternative splicing in plants. Proc Natl Acad Sci U S A 103: 7175–7180
12. Lewis BP, Green RE, Brenner SE (2003) Evidence for the widespread coupling of alternative splicing and nonsense-mediated mRNA decay in humans. Proc Natl Acad Sci U S A 100:189–192
13. Lareau LF, Brooks AN, Soergel DAW, Meng Q, Brenner SE (2007) The coupling of alternative splicing and nonsense-mediated mRNA decay. Adv Exp Med Biol 623:190–211
14. Faustino NA, Cooper TA (2003) Pre-mRNA splicing and human disease. Genes Dev 17: 419–437
15. Cuperlovic-Culf M, Belacel N, Culf AS, Ouellette RJ (2006) Data analysis of alternative splicing microarrays. Drug Discov Today 11: 983–990
16. Ner-Gaon H, Fluhr R (2006) Whole-genome microarray in Arabidopsis facilitates global analysis of retained introns. DNA Res 13:111–121
17. Ner-Gaon H, Halachmi R, Savaldi-Goldstein S, Rubin E, Ophir R, Fluhr R (2004) Intron retention is a major phenomenon in alternative splicing in Arabidopsis. Plant J 39:877–885
18. Modrek B, Lee C (2002) A genomic view of alternative splicing. Nat Genet 30:13–19
19. Royce TE, Rozowsky JS, Bertone P, Samanta M, Stolc V, Weissman S et al (2005) Issues in the analysis of oligonucleotide tiling microarrays for transcript mapping. Trends Genet 21:466–475
20. Laubinger S, Zeller G, Henz SR, Sachsenberg T, Widmer CK, Naouar N et al (2008) At-TAX: a whole genome tiling array resource for developmental expression analysis and transcript identification in Arabidopsis thaliana. Genome Biology 9:R112
21. Zeller G, Henz SR, Widmer CK, Sachsenberg T, Rätsch G, Weigel D, Laubinger S (2009) Stress-induced changes in the Arabidopsis thaliana transcriptome analyzed using whole-genome tiling arrays. Plant j 58:1068–1082
22. Fletcher JC, Brand U, Running MP, Simon R, Meyerowitz EM (1999) Signaling of cell fate decisions by CLAVATA3 in Arabidopsis shoot meristems. Science 283:1911–1914
23. Borevitz JO, Liang D, Plouffe D, Chang HS, Zhu T, Weigel D et al (2003) Large-scale identification of single-feature polymorphisms in complex genomes. Genome Res 13:513–523
24. Bolstad BM, Irizarry RA, Astrand M, Speed TP (2003) A comparison of normalization methods for high density oligonucleotide array data based on variance and bias. Bioinformatics 19:185–193
25. Clark RM, Schweikert G, Toomajian C, Ossowski S, Zeller G, Shinn P et al (2007) Common sequence polymorphisms shaping genetic diversity in Arabidopsis thaliana. Science 317:338–342
26. Zeller G, Henz SR, Laubinger S, Weigel D, Rätsch G (2008) Transcript normalization and segmentation of tiling array data. Pac Symp Biocomput 2008:527–538
27. Swarbreck D, Wilks C, Lamesch P, Berardini TZ, Garcia-Hernandez M, Foerster H et al (2008) The Arabidopsis Information Resource (TAIR): gene structure and function annotation. Nucleic Acids Res 36:D1009–D1014
28. Eichner J, Zeller G, Laubinger S, Rätsch G (2011) Support vector machines-based identification of alternative splicing in Arabidopsis thaliana from whole-genome tiling arrays. BMC Bioinformatics 12:55
29. Sakurai T, Satou M, Akiyama K, Iida K, Seki M, Kuromori T et al (2005) RARGE: a large-scale database of RIKEN Arabidopsis resources ranging from transcriptome to phenome. Nucleic Acids Res 33:D647–D650
30. Boguski MS, Lowe TM, Tolstoshev CM (1993) dbEST-database for expressed sequence tags. Nat Genet 4:332–333
31. Kent WJ (2002) BLAT-the BLAST-like alignment tool. Genome Res 12:656–664
32. Brett D, Pospisil H, Valcárcel J, Reich J, Bork P (2002) Alternative splicing and genome complexity. Nat Genet 30:29–30
33. Simpson CG, Fuller J, Maronova M, Kalyna M, Davidson D, McNicol J, Barta A, Brown JWS (2008) Monitoring changes in alternative precursor messenger RNA splicing in multiple gene transcripts. Plant J 53:1035–1048
34. Sonnenburg S, Schweikert G, Philips P, Behr J, Rätsch G (2007) Accurate splice site prediction using support vector machines. BMC Bioinformatics 8(Suppl 10):S7

Chapter 11

Analysis of In Vivo Occupancy of Aebp1, a Transcription Factor, Using High Resolution Tiling Array

Swati Sinha, Sainitin Donakonda, and Satyanarayana MR Rao

Abstract

Aebp1 is a transcription factor that negatively regulates adipogenesis and also is upregulated in primary GBM as against secondary GBM. Here we describe a systematic approach to map the in vivo occupancy sites of Aebp1 by using a combination of chromatin immune-precipitation technique and hybridization to Agilent high resolution human promoter tiling array (2 × 224) containing 474393 probes. We also describe the use of various bioinformatics tools for analysis of the hybridization data and identification of Aebp1 occupancy sites in U87MG glioma cell line.

Key words Aebp1, Chromatin preparation, Immunoprecipitation, Hybridization to tiling array, Neighborhood analysis, False discovery rate analysis, Motif enrichment analysis, Sequence logos

1 Introduction

Chromatin templated functions and gene expression, in particular, in eukaryotic cells are mainly guided by sequence-specific DNA–protein interactions. Many of the transcription factors also play important role in contributing to a particular disease phenotype. They constitute a considerable number of oncogenes and tumor suppressor genes which control the expression of multiple gene sets associated with tumor development and progression. The key initial event in transcription factor-mediated regulation is the recognition of and binding to specific DNA sequences within the upstream promoter region of a particular gene. These sequences, which are normally of 6–10 nucleotides long, are present across the whole genome. It becomes necessary to identify the occupancy pattern of each of the transcription factor within the genome and also define which of these binding sites are functionally important. Genome-wide predictions by using computational approaches yield much higher number of binding sites than the functionally relevant sites; many of which may also be occluded within the nucleosome. The use of traditional experiments like gel retardation

Tin-Lap Lee and Alfred Chun Shui Luk (eds.), *Tiling Arrays: Methods and Protocols*, Methods in Molecular Biology, vol. 1067, DOI 10.1007/978-1-62703-607-8_11,

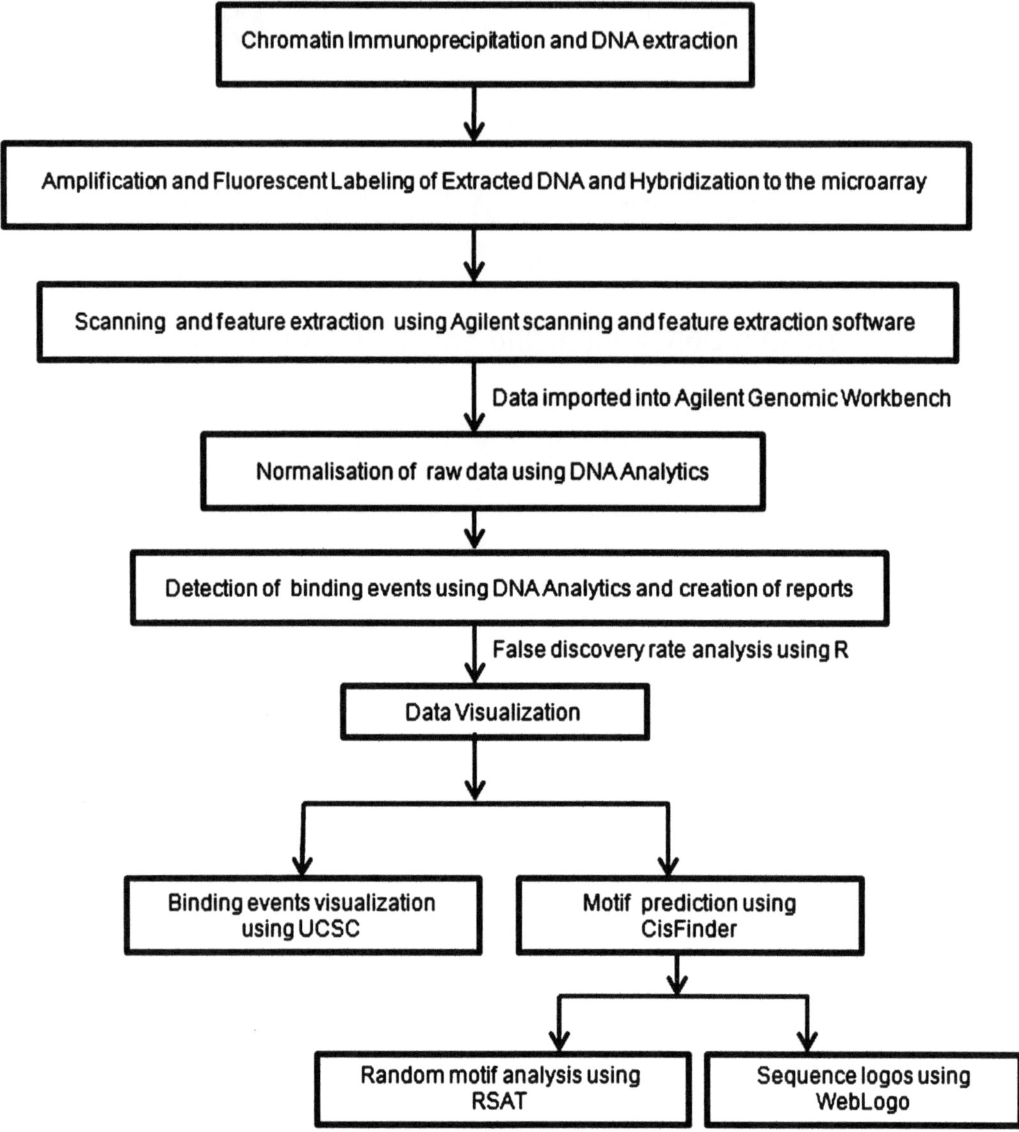

Fig. 1 Workflow for analysis of in vivo occupancy of Aebp1 by Tiling Array

assay (in vitro) and luciferase reporter assay (in vivo) limits the potential binding only to a particular gene in question. High-throughput tools which allow genome-wide studies such as ChIP-chip using tiling arrays are very useful for understanding the basic biological function of DNA binding proteins in genes. Whole Genome tiling arrays can be used for an unbiased, genome-wide occupancy in vivo of a transcription factor. The development of tiling array, a high-throughput technology, has now made it possible to actually understand genome-wide functions of a transcription factor.

Aebp1 (Adipocyte enhancer binding protein 1) was originally identified as a transcriptional repressor that binds to adipocyte enhancer 1 (AE-1 element) located in the proximal promoter region of the adipose P2 gene, which codes for adipocyte specific fatty acid binding protein 4 (FABP4) [1] AEBP1 expression is found to be upregulated in primary GBMs as opposed to progressive secondary GBMs [2]. Since Aebp1 is a transcription factor, it was essential for us to identify the genomic targets to understand the biology of this transcription factor in the context of the disease phenotype. We have recently carried out a detailed study towards this objective [3].

Schematic representation of the workflow undertaken to analyze the in vivo occupancy of Aebp1 (Fig. 1).

2 Materials

2.1 Cell Culture, Cross-Linking and Chromatin Preparation

1. U87MG cells (ATCC, USA).
2. Eagle's Minimum Essential Medium (EMEM) (SAFC Biosciences, Kansas, USA) supplemented with 10 % fetal bovine serum (Gibco by Life Technologies, MD, USA) and 1 % Penicillin–Streptomycin (Sigma, MO, USA).
3. Solution of trypsin (0.25 %) and ethylenediamine tetra acetic acid (EDTA) (1 mM), 1 % formaldehyde (Stock solution of formaldehyde, 36.5 % in H_2O).
4. 0.125 M Glycine for quenching the cross-linking reaction.
5. Cell lysis buffer containing 0.1 % SDS, 0.5 % Triton X-100, 20 mM Tris–HCl pH 8.1, 150 mM NaCl, ImM PMSF, and mammalian Protease Inhibitor (PI) cocktail (Sigma, St Louis, MO). Store at 4 °C and add PMSF and PI at the time of use.

2.2 Checking Sonication Efficiency

1. RNase A (Sigma, St Louis, MO): a 10 mg/ml stock solution in 10 mM sodium acetate buffer, pH 5.2 made by heating to 100 °C for 15 min, cooled to room temperature, and pH adjusted to 7.4 using 0.1 volume of 1 M Tris–HCl, pH 7.4. Aliquots stored at −20 °C.
2. Proteinase K (Sigma, MO, USA) is dissolved at 10 mg/ml in water and aliquots stored at −20 °C.

2.3 Blocking, Preclearing and Immunoprecipitation

1. Blocking buffer containing 10 mM Tris–HCl pH 8.1, 1 mM EDTA, 0.5 % BSA.
2. Protein A agarose beads (Invitrogen, CA, USA).
3. Aebp-1 antibody raised in rabbit (Santa Cruz, CA, USA).
4. IP wash buffers containing 0.1 % SDS, 0.5 % Triton X-100, 20 mM Tris–HCl pH 8.1, 150 mM NaCl, 1 mM PMSF, and mammalian Protease inhibitor cocktail.

5. IP wash buffer containing 0.25 M LiCl, 1 % NP-40, 1 % Sodium deoxycholate, 1 mM EDTA (pH 8.0), 10 mM Tris–HCl, pH 8.1.

2.4 DNA Elution

1. Wash buffer containing 10 mM Tris–HCl pH 8.1 and 1 mM EDTA.
2. DNA elution buffer containing 1 % SDS, 100 mM $NaHCO_3$.
3. RNase A (Sigma, St Louis, MO): a 10 mg/ml stock solution in 10 mM sodium acetate buffer, pH 5.2 made by heating to 100 °C for 15 min, cooled to room temperature, and pH adjusted to 7.4 using 0.1 volume of 1 M Tris–HCl, pH 7.4. Aliquots stored at −20 °C.
4. Proteinase K (Sigma, MO, USA) is dissolved at 10 mg/ml in water and aliquots stored at −20 °C.
5. Phenol solution equilibrated with 10 mM Tris–HCl pH 8.0 and 1 mM EDTA (Sigma, St Louis, MO) and chloroform.
6. Ethanol.
7. 3M sodium acetate pH 5.2, glycogen 25 μg.
8. Nuclease-free water.

2.5 Preparation of Linker, Sample DNA and Amplification for Ligation Mediated (LM) PCR

1. 100 μM of the Oligos: 5′ GCGGTGACCCGGGAGATCT GAATTC 3′ (LMP 1) and 5′ GAATTCAGATC 3′ (LMP 2) was dissolved in nuclease-free water and stored in −20 °C.
2. A mixture of 10 μl each of dATP, dCTP, dGTP, and dTTP (NEB, MA, USA) and 60 μl of nuclease-free water. This was further aliquoted to 25 μl working stocks and stored in −20 °C.
3. T4 DNA polymerase, NEB buffer 2, BSA (NEB, MA, USA).
4. Ligase buffer (Life technologies, MD, USA), T4 DNA ligase (NEB, MA, USA).
5. Thermopol buffer (NEB, MA, USA), Taq DNA Polymerase (Life technologies, CA, USA).

2.6 Fluorescent Labeling and Cleanup of Labeled Genomic DNA

1. Agilent genomic DNA labeling kit PLUS (CA, USA).
2. 0.5–30 kDa Filter.

2.7 Hybridization to the Microarray

1. Human Cot-1 DNA (Life Technologies, CA, USA).

2.8 Scanning Tiling Array Slides and Feature Extraction

1. Agilent scanning and agilent feature extraction software version 10.5 or later.

2.9 ChIP-Chip Data Analysis

1. ChIP interactive Analysis—Agilent Genomic Workbench version 5.0 or later.

2. R statistical software available at http://www.r-project.org/. version 2.15.1 or later.

2.10 Binding Site Visualization, Identification, and Random Motif Analysis

1. Binding site data can be visualized using UCSC genome browser available at http://genome.ucsc.edu/.
2. Motif enrichment analysis can be done using CisFinder software available at http://lgsun.grc.nia.nih.gov/CisFinder/.
3. Binding site sequences can be represented using WebLogo software at http://weblogo.berkeley.edu/.
4. Random sequences can be generated using RSAT tools available at http://rsat.ulb.ac.be/.

3 Methods

3.1 Cell Culture and Cross-Linking of Adherent Cells

1. Grow U87MG cells to 85–90 % confluence in six to eight 140 mm culture plates.
2. Harvest the cells by trypsinization in 0.25 % trypsin–EDTA.
3. Flush the cells using the 35 ml spent medium.
4. Add 1 ml of formaldehyde solution (36.5 %); incubate at room temperature for 10–12 min with intermittent shaking (*see* **Note 1**).
5. Add 5 ml of 1.25 M glycine and incubate at room temperature for 5 min with shaking.
6. Centrifuge at 950 rcf for 5 min. Discard the supernatant and wash the pellet with 1× PBS twice. Centrifuge at 950 rcf for 4–5 min.
7. Drain all the PBS on paper towel and freeze at −80 °C.

3.2 Cell Lysis and Chromatin Preparation

1. Wash the pellet with ice cold 1× PBS. Resuspend the pellet with 0.6 ml cell lysis buffer. Keep for 10 min in ice.
2. Sonicate the cells using Bioruptor [Optimum Pulse—30″ ON & 30″ OFF. No of cycles: 35] (*see* **Note 2**).

3.3 Checking Sonication Efficiency

1. Reverse Cross-linking: Take around 400 μl of the sonicated mixture and incubate at 65 °C overnight.
2. Add RNase A (100 μg/ml) and incubate at 42 °C for 2 h.
3. Add Proteinase K (50 μg/ml) and incubate at its activation temperature for 2 h.
4. Add equal volumes of Phenol–Chloroform (1:1) to the tubes. [Use Tris-saturated phenol pH 8].
5. Vortex vigorously and centrifuge at 16,000 rcf for 3 min.
6. Take the supernatant (aqueous layer) in a fresh new tube, leaving the interphase and the organic layer.

7. Add equal amount of Chloroform–Isoamyl alcohol (1:1), mix and centrifuge at 16,000 rcf for 4 min.
8. Collect the supernatant in new tubes, add 1/10th the volume of 3 M sodium acetate and twice the volume of ethanol (70–75 %) with 30 μg of glycogen.
9. Keep it at −20 °C for 1 h and centrifuge at 16,000 rcf for 10 min.
10. Give 70–75 % ethanol wash and air dry and dissolve in 20 μl of sterile water/TE.
11. Quantify and run on 1–1.2 % Agarose gel.
12. The size of DNA fragments should be between 500 and 800 bp.

3.4 Preclearing (in Duplicates IP1 and IP2)

1. Equilibrate 100 μl (50 % slurry) of protein A agarose beads using 1 ml of cell lysis buffer.
2. Centrifuge at 300 rcf for 3 min. Discard 800 μl of the supernatant. Add 100 μl each from the remaining 200 μl of cell lysis buffer containing beads to the tubes containing cell lysate (*see* **Note 3**).
3. Incubate in an end to end mixer at 4 °C for 1–2 h. Centrifuge at 300 rcf for 3–4 min at 4 °C.
4. Take the supernatant in a fresh tube, add around 8 μg of the antibody to each tube and incubate in end to end mixer at 4 °C overnight (*see* **Note 4**).

3.5 Chromatin Immunoprecipitation

1. Equilibrate 150–200 μl (50 % slurry) of Protein A agarose beads using 1.5 ml of cell lysis buffer. Centrifuge at 300 rcf for 3 min. Discard the supernatant (1,100 μl) and add 200 μl containing beads to IP1 and IP2. Add PMSF to a final concentration 1 mM.
2. Incubate in an end to end mixer for 2–3 h at room temperature.
3. Centrifuge at 300 rcf for 3–4 min. Discard the supernatant and add around 800 μl to 1 ml of the following and wash sequentially with the following buffers.
   - Lysis buffer.
   - Lithium Chloride Buffer + 1 mM PMSF, incubate for 1–2 min in end–end mixer.
   - T.E, incubate for 1–2 min in end–end mixer (repeat the step).
4. Remove solution and add 200 μl of the Elution buffer, incubate at 65 °C for 30 min (gentle intermittent mixing). Centrifuge at 300 rcf at room temperature for 2–3 min and collect the supernatant in a fresh tube. Repeat elution once more (*see* **Note 5**).

3.6 Reverse Cross-Linking and Phenol Chloroform Extraction of Immune-Precipitated DNA

As described previously in Subheading 3.3. Record the concentration of both whole cell extract (WCE) DNA samples.

3.7 Ligation Mediated PCR of IP and WCE DNA

1. Prepare linker DNA by mixing each of LMP1 and 2 (100 μM stock) to a final concentration of 15 μM along with Tris–HCl (1 M stock) pH 7.9 at 0.25 M final concentration. Make 500 μl of this reaction mixture.

   Dispense 100 μl into fresh tubes and run them in thermal cycler in the following program.

I	95 °C	5 min
II	70 °C	1 min
III	Ramp down to 4 °C	0.4 °C/min
IV	4 °C	hold

   Store the linkers at –20 °C till further use.

2. For creating blunt end DNA, take 200 μg of WCE (in duplicates WCE1 and WCE2) and make up the volume to 55 μl.
   (a) 1× NE buffer 2.
   (b) 5 μg BSA.
   (c) 100 μM dNTP mix.
   (d) 1.5 U T4 DNA polymerase.
   (e) Nuclease-free water to make up the volume to each of the DNA samples.

      Cool at 12 °C for 20 min and precipitate the DNA with 11 μl of 3 M sodium acetate (pH 5.2, 1/10th the reaction volume) in the presence of 10 μg of glycogen in each tube.

      Extract DNA by Phenol–chloroform–iso amyl alcohol as described previously.

      Resuspend each of the DNA pellet in 25 μl of nuclease-free water.

      Similarly take 55 μl of ChIP samples (IP1 and IP2) in a separate tube.

      Add 55 μl of Mix containing.

3. To 25 μl of the DNA samples add 25 μl of Ligase mix containing.
   (a) 1× ligase buffer.
   (b) 2 μM linkers (from step 1).
   (c) 200 U of T4 DNA ligase.

(d) Nuclease-free water to make up the volume.

Incubate the reaction mixture at 16 °C for 16 h in a water bath followed by DNA precipitation using 5 μl of 3 M sodium acetate (pH 5.2, 1/10th the reaction volume).

Precipitated DNA is resuspended in 25 μl of Nuclease-free water to make up the volume.

3.8 Amplification of DNA Samples

1. To each of the PCR tubes containing 25 μl of IP1, IP2 and WCE1, WCE2 add 15 μl the following mixture containing.
   (a) 1× Thermopol Buffer (NEB).
   (b) 250 μM dNTP mix.
   (c) 1 μM LMP1 oligonucleotide.
   (d) Nuclease-free water to make up the volume.
2. Prepare 10 μl for each reaction of the following (Polymerase Mix).
   (a) 1× Thermopol Buffer (NEB).
   (b) Taq polymerase 0.25 U.
   (c) Nuclease-free water to make up the volume.
3. Run a Ligation-mediated PCR cycle with the following program.

I	55 °C	4 min. Hold the reaction at 2.0 s and add 10 μl of the polymerase mix at 55 °C and continue
II	72 °C	3 min
III	95 °C	2 min
IV	95 °C	30 s
V	60 °C	30 s
VI	72 °C	1 min
VII	Repeat **steps IV** to **VI**	×14 times
VIII	72 °C	5 min
IX	4 °C	Hold

4. Transfer the contents to a 1.5 ml microfuge and add 475 μl of nuclease-free water. Pipette out 5 μl of this diluted PCR product and store the remaining in −20 °C.
5. To the diluted PCR product add 45 μl of the following.
   (a) 1× Thermopol Buffer (NEB).
   (b) 250 μM dNTP mix.
   (c) 1 μM LMP1 oligonucleotide.
   (d) Taq polymerase 1.25 U.
   (e) Nuclease-free water to make up the volume.

6. Run the following PCR program.

I	95 °C	2 min
II	95 °C	30 s
III	60 °C	30 s
IV	72 °C	1 min
V	Repeat **steps II** to **IV**	×24 times
VI	72 °C	5 min
VII	4 °C	Hold

7. Precipitate the PCR product using 625 mM ammonium acetate and resuspend the pellets to a final concentration of 100 ηg/μl.

3.9 Fluorescent Labeling of DNA

1. To 2 μg of each of WCE1, WCE2 a1nd IP1, IP2 add nuclease-free water to make up the volume to 26 μl.
2. Add 5 μl of random primers (Agilent Genomic DNA Labeling Kit PLUS) and mix by gentle tapping. Spin down and incubate at 95 °C for 3 min on a thermal cycler. Chill on ice for the next 5 min.
3. Prepare labeling mix (Agilent Genomic DNA Labeling Kit PLUS) on ice by adding the following components in the dark.
   (a) 5× buffer.
   (b) 1× dNTP.
   (c) Cyanine 3-dUTP (1 mM) for WCE.
   (d) Cyanine 5-dUTP (1 mM) for IP.
   (e) Exo-Klenow fragment (1 μl).
4. Add 19 μl of this mixture to the rest of the sample kept on ice to make a total of 50 μl. Mix by gentle tapping and spin down. Put the reaction in a thermal cycler at 37 °C for 2 h.

   Stop the reaction by inactivating the enzyme at 65 °C for 10 min and store in −20 °C in dark.

3.10 Cleanup of Labeled Genomic DNA

1. Transfer labeled samples to a 1.5 ml microfuge tube and add 430 μL of TE (pH 8.0) to each reaction tube.
2. For each labeled sample, place a Amicon Ultra −0.5 to 30 kDa filter into a 1.5-ml microfuge tube and load each labeled amplified WCE or IP sample into the filter. Spin 10 min at 16,500 rcf at room temperature. Discard the flow-through.
3. Add 480 μL of TE (pH 8.0) to each filter. Spin for 10 min at 16,500 rcf at room temperature. Discard the flow-through.

4. Invert the filter onto a 1.5 ml microfuge and spin at 1,000 rcf to collect the purified sample. If the sample exceeds 41 μl then repeat **step 2** and if sample volume is less than 41 μl, make up the volume with TE (pH 8).
5. Repeat the process to collect a total of 79 μl of mixture volume.
6. Take 1.5 μL of each sample to determine yield and specific activity.

   Concentration of DNA > 5 μg.

   Cyanine 5 > 2 pmol/μl.

   Cyanine 3 > 3 pmol/μl.

3.11 Hybridization to the Microarray

1. Prepare the 10× Blocking Agent (Agilent Oligo aCGH Hybridization Kit) by reconstituting in 1.35 ml of nuclease-free water. Incubate at room temperature for 60 min to dissolve completely. Check for undissolved particles before proceeding.
2. Set heat blocks at 95 °C and water baths at 37 °C.
3. Combine Cy5- and Cy3-labeled samples with nuclease-free water in a 1.5 ml tube for a total volume of 79 μl.

   Add the following to the tube.

I	Human Cot-1 DNA (1.0 mg/ml)	25 μg
II	Agilent Blocking Agent (10×)	26 μl
III	Agilent Hybridization Buffer (2×)	130 μl

Tap, mix, and spin down.

Heat sample at 95 °C for 3 min. Transfer to 37 °C water bath immediately and incubate for 30 min.

Collect sample by centrifugation at 16,500 rcf.

Arrange Agilent Microarray Hybridization Chamber as per manufacturer's instructions and load 240 μl of the two individual replicates onto the glass gasket slide.

Mount the Microarray probe slide with the "active side" facing the sample and assemble the chamber.

Sway the chamber vertically to access the presence of any bubble (mobile air bubbles do not interfere with the hybridization process).

Incubate at 65 °C for 40 h in a hybridization chamber at 20 rpm.

3.12 Washing Microarray Slides

1. Pre-warm Oligo aCGH/ChIP-on-chip Wash Buffer 2 (Agilent) at 31 °C.

   Wash Slide rack and staining dishes in Milli-Q water before use.

   Arrange the slide-staining dishes with the magnetic bead on the stirrer with buffers at required temperatures.

2. Fill the first slide-staining dish with Oligo aCGH/ChIP-on-chip Wash Buffer 1 (Agilent) at room temperature. Place a slide rack into the second slide-staining dish. Fill with Oligo a CGH/ChIP-on-chip Wash Buffer 1 to completely submerge the slide rack.
3. Disassemble the Agilent Microarray Hybridization Chamber containing the microarray slide. Minimize exposure to air. Completely submerge the gasket–microarray slides into the first slide-staining dish. The gasket glass slide should be down and the microarray slide numeric barcode should face up.

   Separate the two with a pair of forceps and let the glass gasket fall to the bottom of the dish.
4. Quickly transfer the slide into the slide rack in the second-staining dish containing Oligo aCGH/ChIP-on-chip Wash Buffer 1 at room temperature.
5. Repeat for the second slide.
6. Transfer the Oligo aCGH/ChIP-on-chip Wash Buffer 2 onto a third slide-staining dish. Maintain the temperature at 31 °C.
7. After washing both slides in Oligo aCGH/ChIP-on-chip Wash Buffer 1 for 4–5 min under constant stirring transfer the slide rack to Oligo aCGH/ChIP-on-chip Wash Buffer 2 containing slide-staining dish. Wash for 1 min.
8. In the fume hood, fill the fourth slide-staining dish with acetonitrile and the last slide-staining dish with Stabilization and Drying Solution (Agilent).
9. Immediately transfer the slide rack to the fourth slide-staining dish containing acetonitrile and stir using setting 4 for 10 s.
10. Transfer slide rack to the last slide-staining dish filled with pre-warmed (37–40 °C) Stabilization and Drying Solution, and stir using setting 4 for 30 s.
11. Slowly remove the slide rack trying to minimize droplets on the slides.
12. Scan the slides immediately to minimize impact of environmental oxidants on signal intensities.

3.13 Scanning Tiling Array Slides

1. Take hybridized slides and place slide in the holder such that the numeric barcode is visible. Assemble slide holders into carousel. Do not place any slide in slot marked "H" and verify scan settings for two-color scans.
2. To alter settings, click Settings > Modify Default Settings. A window pops up in which you can change the settings.
3. Select settings for the automatic file naming.
   - Prefix 1 is set to Instrument Serial Number.
   - Prefix 2 is set to Array Barcode.

4. Once the scanner is ready, click Scan Slot *m* – *n* on the Scan control main window, where "*m*" represents the start slot for the first slide and "*n*" represents the end slot for the last slide.

3.14 Data Extraction Using Feature Extraction Software

Probe features from microarray scan data using feature extraction (FE) software which allows measuring gene expression in the experiment. Most recent feature extraction software can be downloaded from www.agilent.com/chem/fe. Extract .tif images using this feature extraction software from microarray scan images.

1. Open the Agilent Feature extraction software. Recent Feature extraction protocols for gene expression can be found at www.agilent.com/chem/FEprotocols.
2. Add the images (.tif) to be extracted to the FE project. Click on Add New Extraction Set icon on the toolbar and browse the location of the .tif files, select the .tif files, and click Open.
3. The Feature Extraction program (FE) automatically assigns a default grid template and protocol for each extraction set. To access the FE Grid Template properties, double click on the grid template in the Grid Template Browser.
4. Set FE Project Properties: Select the Project Properties tab and in the General section enter name in operator text box. In the Input section verify that at least the following default settings as shown in the Agilent ChIP-chip protocol manual [4].
5. Check the Extraction Set Configuration.
   - Select the Extraction Set Configuration tab.
   - Verify that the correct grid template is assigned to each extraction set in the Grid Name column. To assign a different grid template to an extraction set, select one from the pull down menu.

     Grid Template Browse can be utilized to add any grid templates. To add, right-click inside the Grid Template Browser, select Add and browse for the design file (.xml) and click Open to load grid template into FE database.

   Latest grid templates can be updated online. Right-click Grid Template Browser and select Online Update. Most recent grid templates are available from Agilent Website at www.agilent.com/chem/downloaddesignfiles. After downloading grid templates, it should be added to Grid Template Browser (*see* **Note 6**).
6. Verify that the protocol name (ChIP-v1_95_May07) is assigned to each extraction set. Using drop down menu, different protocol can be assigned to an extraction set. If a protocol is not available, FE protocol browser can be imported via FE protocol browser. To import, right-click FE protocol Browser,

select import. Browse for the FE protocol (.xml) and click Open to load the protocol into the FE database. Latest protocols can be downloaded from www.agilent.com/chem/FEprotocols.

7. Select File > Save As to save FE project (.fep) and check icons for image files in the FE project (window no longer should have red X through them). Red X through indicates that extraction protocol was not selected. Reselect the extraction protocol for the same image file.
8. Select Project > Start Extraction. After extraction is completed successfully, view Quality Control report for each extraction set by double clicking the QC report link in summary report tab. Also verify that the grid has been properly placed by inspecting spots at the four corners of the array.

3.15 Importing Data to Agilent Genomic Work Bench

The text output files generated via Images using Agilent feature extraction software were used for the analysis.

1. Open Agilent Genomics Workbench software and select Home tab, Click Import > Array Files > FE File. Select a file for import; click on its name and additional files control-click their names.
2. Next click open to import Agilent FE files, in either case the Agilent Extraction Import dialog box will appear. In this box please make settings as required (as shown in agilent ChIP interactive analysis manual) [5].
3. Then do one of the following:
   (a) To import the files by clicking OK.
   (b) To import the files in the background, click Run in back ground. This allows you to work while the program imports the files. Usually, the program uses the genome build specified in the array design file and protects it from changes.
4. To import array-specific attributes Click File > Import > Array Attributes. Select the desired microarray attributes file and click Import.
5. Import gene list via My Entities List pane, double click the Entities folder to expand it. Right-click the Gene List folder, then click Import Gene List. An Import dialog box appears.
6. Export gene list via My Entity List pane, in Gene List folder, right-click the gene list export, then click > Save As > type in filename > click save > OK.
7. Export tracks as a BED format track file using Home tab, click Export > Tracks. Mark the tracks that you want to export or Select All and click OK. Select a location and type a name for the exported track file, and then click Export. Finally use genome browsers like UCSC.

3.16 Normalization of Tiling Arrays Using DNA Analytics

Understanding true variation from binding event is vital to answer the biological questions that the ChIP experiment was designed to address. In order to do this normalization techniques must be applied which allows us to limit the systematic variation in the data such that true biological variations are detected. DNA Analytics software (part of Agilent genomic workbench) is used to perform normalization procedures for both intra- and inter-array normalization as follows.

1. In the preprocessing tab click > Normalization, select User Defined. The Edit button becomes available.
2. Click > Edit button, a dialog box with available normalization methods will appear as follows.
   (a) *Median Blank subtractions.* For each channel (IP or WCE) of each replicate of each array, the median of the intensities of the blank spots are subtracted for each on that replicate. As in all normalization steps, any negative signal intensities that result will cause the probe to be flagged as "excluded."
   (b) *Inter-array median normalization.* This normalization step adjusts the signals so that each IP or WCE channel has a similar distribution on each replicate of each array. For each replicate of each array, the median signal over the common probes is computed separately for each channel. Then, for each channel, the arithmetic average of this median is computed over all replicates of all arrays.
   (c) *Dye-bias median normalization.* This normalization step corrects for dye bias. Only signals in the IP channel are affected. The signal in the IP channel is multiplied by a factor that is computed for each replicate of each array.
3. Mark the desired normalization methods under select normalization. To set specific parameters of a normalization method, click the name of the method and click OK [5].

3.17 Detecting Protein–DNA Binding Events

The protein–DNA binding event detection algorithm used in DNA Analytics is the Whitehead Per Array Neighborhood Model. This method uses the distribution of all probes on each array to compute regions of increased probe signal (termed "peaks"). The algorithm does this by examining groups of probe triplets which are significantly enriched, yielding robust binding event detection. The Whitehead model samples every probe and its immediate upstream or downstream neighboring probe to identify a robust estimate of the location of bound protein [6].

3.17.1 Configuring an Error Model

1. In the Preprocessing tab, under Error Model, select > Whitehead Error Model.
2. To set parameters for the Whitehead error model select Edit button under Error Model parameter settings dialog box

appears. Then set parameters as shown in agilent ChIP interactive analysis manual [5].

3. Click Ok (*see* **Note 7**).

3.17.2 Apply Event Detection to the Current Experiment

1. After the error model configured click > Analysis under Event Detection, click the Edit button. Then the Whitehead per Array neighborhoods model parameter settings dialog box appears.
2. The parameters are described of the three-probe model shown in table (agilent ChIP interactive analysis manual). And Click OK.
3. Under Event Detection, mark apply which in turn applies event detection to the experiment.

3.18 Creating Reports

Finally create various reports as probe report, gene report, and quality report.

3.18.1 Probe Report

The ChIP application makes information about the probes in the current experimental result available in tab-separated value (*.tsv) format. A Probe report contains one row for each probe in the array. Separate files for each array will be generated. These reports can be viewed in any spreadsheet program.

To create a probe report on the current experimental results.

1. Click Reports > Probe Report then a parameter settings dialog box appears.
2. Click > Browse select a folder for the report and type the name in appropriate folder.
3. Click > Open then location of the report appears in the probe report settings along with report settings dialog box in report location and Click > OK.
4. Finally these steps will create probe report in the format as shown in agilent ChIP interactive analysis manual [5].

3.18.2 Gene Report

The ChIP application creates information about genes in current experimental result available in tab-separated value (.tsv) format. This gene report contains one row for probe in the array, grouped by the genes to which the probes bind. It also includes loci represented by probes on the array that are not associated the genes.

1. Click Reports > Gene report. Then gene report settings parameter settings dialog box appears as shown in agilent ChIP interactive analysis manual.
2. Check boxes depending on the requirement. For example, user can check > show only gene names vice versa.
3. Click Browse > open appropriate folder for report and name the file.

4. Click > Open the path name of the report appears in parameter settings dialog box.
5. Click OK. This analysis will create the gene report has information showed in agilent ChIP interactive analysis manual [5]. To view the report file, any spread sheet program can be used.

3.18.3 Quality Report

The QC report summarizes the current analysis, and the overall statistics of each array. In addition to summary tables, it includes plots that summarize the data graphically. The program creates the QC report in HTML format.

1. Click Reports > QC Report. Then QC report settings parameter settings dialog box appears.
2. Click Browse > select appropriate name.
3. Open a folder for the report and type in file name.
4. Click > Open the location of the report appears in the QC report settings parameter settings dialog box.
5. Click OK. The program creates a folder that contains the report files. A dialog box appears.
6. To display a report, follow these steps: (a) Click Yes. The QC report folder will be created (b) Open the desired folder and click for QC report as HTML file and view it in internet browser.

3.19 Statistical and False Discovery Rate Analysis

Binding events detected using Whitehead Per Array Neighborhood Model analysis reveals enriched genes to which target transcription factor will be bound. Based on *p*-value we can derive statistical significant genes. Further to minimize any false positives all enriched genes should be subjected to false discover rate analysis using R software [6] as follows.

1. Take *p*-values from tiling array analysis.
2. Open R software console.
3. Use p.adjust package in R console and import *p*-value data (set at 0.05 by user) and apply Benjamini & Hochberg method [7].

3.20 Data Visualization

Besides binding event detection binding site analysis is vital in ChIP-chip analysis. To understand results from the data it is essential to visualize data using genome browsers. These browsers are powerful tools to visualize ChIP-chip data as a track against an annotated genome. UCSC genome browser is a web-based genome browser available at http://genome.ucsc.edu/ [8]. This contains annotation tracks that are computed at UCSC from publicly available data and provided by collaborators. One can view the data along the chromosome by zooming in or providing genomic coordinates.

3.21 Binding Site Analysis and Visualization

Protein–DNA binding site analysis is central in ChIP-chip analysis and it is essential to characterize potential binding site(s) where target transcription factor is bound throughout the genome. The CisFinder an web application which identifies transcription factor binding sites from ChIP-Chip data is available at http://lgsun.grc.nia.nih.gov/CisFinder/ [9].

1. Open CisFinder software by clicking http://lgsun.grc.nia.nih.gov/CisFinder/.
2. Create FASTA format of probe sequences from ChIP-chip analysis (*see* **Note 8**).
3. User can create login id for software via registering in the website.
4. Once user logged in, input page will open. Load probe sequences (FASTA format).
5. Then scroll down to select data files there are several pull down menus in that go to test menu and choose ChIP-chip dataset.
6. Click identify motifs in analysis tool section then parameters can be defined to identify motifs.
7. Finally list of enriched motifs along with their position weight matrices list can be visualized.

3.22 Random Binding Site Motif Analysis

It is essential to control the reliability of motif prediction programs such as CisFinder. There are several ways to build negative control sets, i.e. data sets that do not contain significant motifs. This comparative evaluation between discovered motifs with random motifs offers strong evidence about biological significance of motifs. Random-Seq program in RSAT tools (Regulatory Sequence Analysis Tools) available at http://tagc.univ-mrs.fr/rsat/ [10] can be used to build random sequences as follows.

1. Open RSAT tools by clicking http://tagc.univ-mrs.fr/rsat/.
2. In this application go to build control sets and click > random sequence.
3. In general options > specify sequence length and number of sequence required (usually same number of sequences as ChIP-chip data). And finally choose FASTA format.
4. In background model choose organism of interest.
5. Output can be sent displayed on the page or email and Click > GO.
6. Once all random sequences are downloaded in FASTA format, repeat binding site analysis using CisFinder as mentioned above.

3.23 Sequence Logos Representation

Once the binding motifs identified user can represent graphical sequence logo by using WebLogo software available at http://weblogo.berkeley.edu [11].

1. Open WebLogo application by clicking http://weblogo.berkeley.edu/.
2. WebLogo accepts aligned set of potential binding site sequences as input.
3. Click > Create logo.
4. Finally this creates graphic that represents each nucleotide by a color-coded stack letters corresponding to each of the four bases height of each letter represents the proportion of sequence conservation of binding site. Then sequence logo can be exported in many formats including GIF and PNG, as well as print friendly formats such as PDF.

4 Notes

1. Fixation in formaldehyde solution may vary with every cell type and nature of interaction. Cross-linking must be standardized for each DNA–protein interaction. Excessive cross-linking reduces antigen accessibility and sonication efficiency, whereas inappropriate fixation reduces amount of antigen–antibody interactions during chromatin immunoprecipitation. Concentration of formaldehyde solution, time of incubation, and temperature may require optimization.
2. Sonication cycle may differ depending on cell number, cell type, and procedure followed for chromatin isolation. If the fragment size of the DNA is large it may lead to high background and vice versa. Cell number and extent of cross-linking may bring about variations in sonication efficiency.
3. Protein A agarose beads to be used for immunoprecipitation may be blocked with 1 % bovine serum albumin (Sigma) to reduce nonspecific binding.
4. Antibody concentration makes a discernable difference between signals to noise ratio. The optimum antibody concentration varies depending on the stringency of buffer and affinity of the antibody.
5. It is essential to remove all RNA from the reverse cross-linked samples to eliminate possibility of hybridization to the micro array chip in subsequent steps.
6. After adding new grid template to the Grid Template Browser, specify the default protocol for the new grid template to automatically assign an FE protocol to an extraction set.
7. P (Xbar): This parameter refers to the average *P*-value for the central probe and its neighbors. The value set is <0.05 by this user can define statistically significant enriched genes.
8. Take enriched probe sequences from ChIP-chip analysis and import them to Seqret tool in EMBOSS web-based application

available at http://bips.u-strasbg.fr/EMBOSS/. Upload all probe sequences into this tool and convert them into appropriate pearson FASTA format.

Acknowledgments

This work was supported by grants from NMITLI program of Council of Scientific and Industrial Research (CSIR) and Department of Biotechnology (DBT), New Delhi.

References

1. He GP, Muise A, Li AW, Ro HS (1995) A eukaryotic transcriptional repressor with carboxypeptidase activity. Nature 378:92–96
2. Reddy SP, Britto R, Vinnakota K, Aparna H, Sreepathi HK, Thota B et al (2008) Novel glioblastoma markers with diagnostic and prognostic value identified through transcriptome analysis. Clin Cancer Res 14:2978–2987
3. Ladha J, Sinha S, Bhat V, Donakonda S, Rao SM (2012) Identification of genomic targets of transcription factor Aebp1 and its role in survival of glioma cells. Mol Cancer Res 10:1039–1051
4. Agilent Mammalian ChIP-on-chip Protocol Version 10.11, January 2010. www.genomics.agilent.com/files/Manual/G448190010_MammalianProtocol_10.11.pdf. Accessed 9 Oct 2012
5. Agilent ChIP Interactive Analysis user guide Agilent technologies Version 5.0, January2010. www.chem.agilent.com/Library/usermanuals/Public/chip_interactive_analysis.pdf. Accesses 10 Oct 2012
6. R Development Core Team (2009) A language and environment for statistical computing. R Foundation for Statistical Computing, Vienna, Austria
7. Benjamini Y, Hochberg Y (1995) Controlling the false discovery rate: practical and powerful approach to multiple testing. J R Stat Soc Ser B 57:289–300
8. Kent WJ, Sugnet CW, Furey TS, Roskin KM, Pringle TH, Zahler AM et al (2002) The human genome browser at UCSC. Genome Res 12:996–1006
9. Sharov AA, Ko MS (2009) Exhaustive search for over represented DNA sequence motifs with CisFinder. DNA Res 16:261–273
10. Thomas-Chollier M, Defrance M, Medina-Rivera A, Sand O, Herrmann C, Thieffry D et al (2011) RSAT 2011: regulatory sequence analysis tools. Nucleic Acids Res 39: W86–W91
11. Crooks GE, Hon G, Chandonia JM, Brenner SE (2004) WebLogo: a sequence logo generator. Genome Res 14:1188–1190

Chapter 12

Application of the Simple and Efficient Mpeak Modeling in Binding Peak Identification in ChIP-Chip Studies

Ming Zheng, Yunmin Li, and Yun-Fai Chris Lau

Abstract

Chromatin immunoprecipitation and hybridization of high-density promoter microarray (ChIP-chip) is a powerful strategy to identify target genes for specific transcription factors and other DNA-binding nuclear proteins in a genome-wide manner. Services of core facilities have greatly enhanced the accessibility of these technologies to new investigators to the field. The Mpeak modeling is a simple and efficient computer program, capable of identifying chromatin-binding peaks in ChIP-chip datasets. It utilizes advanced statistical computation, but yet offers a simple procedure with user inputs on parameters in its operation. The Mpeak-fitted signals are tabulation in convenient formats and can be visualized in various genome-display graphic programs, including SignalMap and Genome Browser, and analyzed together with other datasets, such as microarray expression patterns. Several research groups have used the Mpeak program in their respective ChIP-chip studies. The various features of Mpeak will be illustrated with ChIP-chip datasets from a study designed to identify the target genes for the sex-determining factor, SRY, in mouse embryonic gonads at the time of sex determination.

Key words Mpeak modeling, Peak identification, ChIP-chip, SRY targets

1 Introduction

Since the completion of the Human Genome Project, in which the entire human genome was totally sequenced [1], rapid advances in genomic technologies have greatly facilitated research on genetics and genomics in many fields [2–11]. Two technologies, the hybridization-based microarray and next-generation sequencing strategies, stand out to be the pivotal foundation for these advances. As evidenced by the presentations in this volume, and many other publications in the scientific literature, these two technologies have offered opportunities to investigate complex biological and biomedical problems in a global manner. However, these new technologies demand significant financial investments on instruments, experienced staffing, computational powers, and knowledge in biostatistics and bioinformatics, which usually are not readily

Tin-Lap Lee and Alfred Chun Shui Luk (eds.), *Tiling Arrays: Methods and Protocols*, Methods in Molecular Biology, vol. 1067, DOI 10.1007/978-1-62703-607-8_12,

available to many laboratories, primarily focusing on basic physiological, morphological, and traditional molecular biology and genetic and biochemical approaches in their investigations. On the other hand, these investigators are likely to be most knowledgeable in their respective problems being studied, and might possess key reagents and/or knowledge, such as transgenic mouse strains, specific/unique antibodies to key molecules, and in-depth understanding of the biological, physiological, and/or developmental processes. Recent availabilities of core facilities among educational and research institutions and for-profit commercial vendors have significantly enhanced the participation of these basic biologists and new investigators in reaping the benefits of the advanced genomic technologies in their respective research topics. The establishment of numerous public online portals, such as the Genome Browser at the University of California, Santa Cruz [12], the Database for Annotation, Visualization and Discovery (DAVID) at the National Institute of Allergy and Infectious Diseases (NIAID) [13], the Galaxy (http://galaxyproject.org/) [14], and the Cistrome (http://cistrome.org/ap/) [15], have further facilitated the downstream analytical aspects of these types of advanced genomic research.

Chromatin immunoprecipitation (ChIP) is a powerful means to determine the specific bindings of transcription factors and nuclear proteins to the genomic chromatin/DNA. Two specific strategies, i.e., ChIP-chip [16] and ChIP-seq [17–19], have been established with the hybridization-based microarray and next-generation sequencing technologies, respectively, in which the binding patterns of a transcription factor/chromatin-binding protein could be determined in a genome-wide manner. Both ChIP-chip and ChIP-seq have offered important investigative strategies in many biological, biomedical, and developmental problems. The identification of putative target genes of a critical transcription factor(s) in a developmental event will provide invaluable insights into gene regulatory hierarchy, pathways, and networks involved in such biological development [13]. Although core facilities for microarray analysis and next-generation sequencing will assist the new investigators to perform the microarray hybridization and next-generation sequencing portions of their ChIP-chip and ChIP-seq studies, the analytical aspects of the projects could still be challenging tasks. Collaborations with experienced biostatisticians in these areas will certainly increase the possibility of success in such endeavors. Availability of reliable strategies and operationally simple computer programs and resources will be essential for these investigators to participate in the analysis and evaluation of the results, so obtained with the advanced genomic technologies. The model-based Mpeak program has been developed to analyze the results of ChIP-chip experiments [20], to identify putative chromatin/DNA-binding peaks with statistics on relative signal strength and confidence, and to generate result files readily visualized with established computer programs, such as SignalMap (Roche-NimbleGen Inc., Madison, WI), and online portals, such as the Genome Browser (University of California, Santa Cruz) for genome-wide graphic display and other downstream analyses of the results. Several research groups have successfully utilized Mpeak to generate consistent and confirmable results, pertaining to their respective topics and biological problems [16, 21–24].

In this chapter, we have outlined the principle of the Mpeak modeling, procedures to execute the program, and steps to incorporate the Mpeak results with other online resources for visualization and interpretation. We have further illustrated these features with a couple of examples, based on our investigation on identifying target genes for the sex-determining factor, SRY, in mouse embryonic gonads at the time of sex determination.

2 Materials

2.1 Mpeak Software

Mpeak is Microsoft Windows-based software used to identify the peaks in the signal data from ChIP-chip experiments. The software and source codes (compiled using Microsoft Visual Studio 2008, http://www.microsoft.com/) for nonprofit research uses can be downloaded from http://labs.medicine.ucsf.edu/chrislau/data-software.html/.

2.2 R Software

The statistical software R (http://www.r-project.org) is a powerful and widely used software to perform statistical analysis as well as many bioinformatics analysis. It is freely available in many platforms.

2.3 Data from Replicated ChIP-Chip Experiments to Be Analyzed

Results from hybridization of promoter tiling microarrays from a variety of vendors could be used in Mpeak modeling for binding peak identification. The results should be presented as log2 ratios between sample and input hybridization intensities and corresponding chromosome coordinates in general feature format (GFF). Please also see Notes section for information on GFF. Most of our experience is derived from Mpeak analyses of ChIP-chip data on human and mouse 2.5-kb promoter tiling microarrays from Roche-NimbeGen Inc.

3 Methods

3.1 Analysis of ChIP-Chip Data Using Mpeak Modeling

3.1.1 Brief Introduction to the Principle of Mpeak Modeling

ChIP procedures normally involve random fragmentation of the chromatin to optimal sizes, as judged by the DNA sizes of the fragmented chromatin, before the immunoprecipitation step. Such process of randomly chopping genomic DNA on either side of a reference, i.e., DNA-binding site, can be modeled by a Poisson process. In Mpeak modeling, it is postulated that the probability of a chop happens between one base pair and the next one is a small constant. Therefore, on each side of the binding peak, the probability of DNA bases to be on the same randomly chopped DNA fragment as the binding site decreases exponentially with increase in distance from the peak. This argument leads to a statistical model for a binding site: the signal at the peak is the highest on the DNA fragment and the signals will decay linearly on both sides (albeit the speed of decay could be different). Mpeak takes advantage of this triangular model and uses linear regression analysis to identify where signals are enriched and to pinpoint the exact peak position. This method has also been extended to study biological replicated datasets. In the presence of multiple ChIP-chip datasets, Mpeak will fit separate triangular model with potentially different range of the triangle to each of the replicates centered at the same peak position only if the peak is present on similar, as defined by parameters below, genomic locations of the promoter tiles on the replicated datasets. The statistical rationale and considerations of the Mpeak modeling have been presented in full details elsewhere [20].

3.1.2 Running Mpeak

A "click-and-play" implementation of the Mpeak program for Microsoft Windows operation systems has been developed (implemented using Microsoft Visual Studio, Microsoft, Seattle, WA, version 9). To use the program, users just need to specify the necessary input and output files and set the parameters for Mpeak.

1. Start the PeakDector program.
2. Under pull-down menu select Analyze, and then Replicated Experiments.
3. Input the number of replicates (input "1" if there is only a single experiment) and then open the replicates one by one. It may take a few minutes to open the GFF files containing the normalized hybridization intensities at the promoter tiles.
4. Adjust the parameters for PeakDetector in the pop-up window (Fig. 1)—see the discussion below for selection criteria. Click OK to accept the parameters.
5. Specify the names and where the Mpeak-fitted peaks, peaks only, and adjusted p-value output files (i.e., output _shape1, output_peaks_only, and adjusted _Pvalue, respectively, for the detected peaks) should be saved, and click OK.

3.1.3 Discussion on Parameter Settings on Mpeak

1. Promoter tiling microarrays from different vendors or different versions from the same vendor could have slightly different GFF format for their output files. The normalized ratios of hybridization intensities between experimental and input (control) samples could be in either linear or log2 format. Since the modeling of Mpeak is based on the log2-based

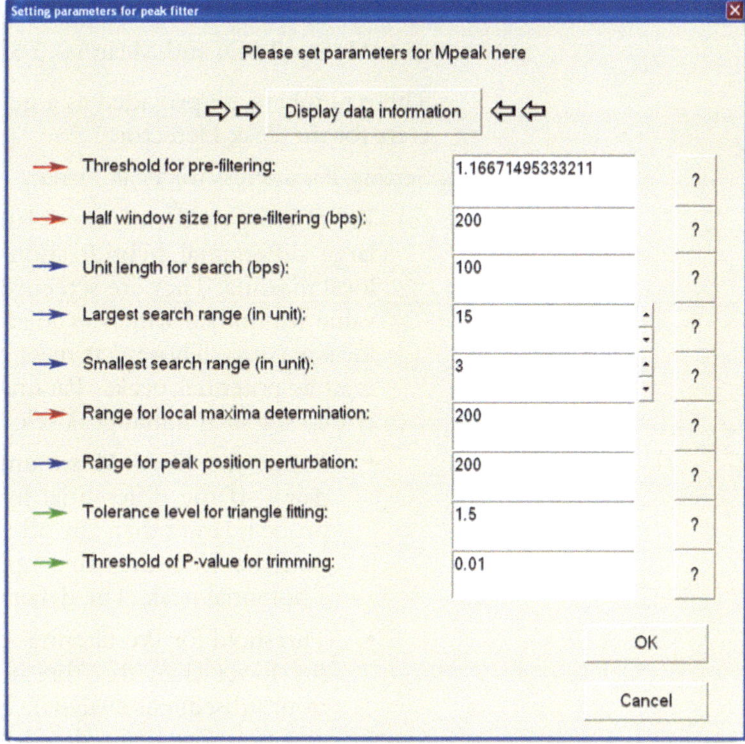

Fig. 1 A screen display of input parameters for Mpeak analysis. A pre-analysis calculation of the means and standard deviation can be viewed by activating the "Display data information" *box* (indicated by *double-hollow arrows*). *Red*, *blue*, and *green arrows* indicate the relevant parameters for pre-filtering, selection for local maxima and *triangle* modeling, and post-selection processing, respectively. The small buttons (with *question marks*) on the *right*, next to each parameter entry row, can be activated to display a brief discussion of the respective parameter

signal intensity, it is necessary to convert the intensities in linear format to log-scale, which can be done easily in Microsoft Excel or R (http://www.R-project.org). Descriptive/comment lines starting with the "#" sign will be automatically recognized by Mpeak and therefore, there is no need to remove such lines (as required in older versions of Mpeak). The detected peaks as well as the fitted peak signals from Mpeak are presented in GFF format, which could be best visualized with a log2 format in either SignalMap or Genome Browser. For more about GFF, please see description at the Welcome Trust Sanger Institute Web site, http://www.sanger.ac.uk/resources/software/gff/.

2. At the "Setting Parameters" screen (Fig. 1), there is a small button, "Display Data Information" (indicated by double-hollow arrows, Fig. 1). Activation of this button will display a small screen showing the statistics of the raw data:

   Mean of Data; Standard Deviation (SD) of the data; Mean + 1 SD.

Mean + 1.5 SD.

Mean + 2 SD; and Mean + 2.5 SD of the data.

These numbers are provided as a guide for setting the parameters for the Peak Detector.

3. Setting Parameters for Peak Fitter.

   (a) Pre-filtering to identify candidate peaks. First, probes with large differential hybridization signals are identified as local maxima. They are screened in a defined pre-filtering value within half windows adjacent on both sides of such local maxima. Those that meet the criteria will be considered as potential peaks. Parameters (Fig. 1, red arrows) critical for such initial peak selection are as follows:

      - Range for Local Maximum Determination (in base pair)—If the differential hybridization intensity of a probe is larger than any other probes within this range on either side, the subject probe will be considered as a potential peak. The default value is 200 bp.
      - Threshold for Pre-filtering—Local maxima with intensity lower than this threshold are filtered (excluded from subsequent analysis). This is the most important parameter, with a predetermined *p*-value limit (below). It defines the stringency of the Mpeak peak fitting function. Setting a low number may include tiny peaks, which might be false positives. Setting a high number may exclude true peaks in the results. The default value is mean + 1 SD of the raw data. The "Display Data Information" function will provide preliminary statistics for the raw data as references. User should explore various values for pre-filtering setting, such as mean + 1 SD and mean + 1.5 SD, or increasing numerals, e.g., 1, 1.5, 2, and 2.5, for evaluation, and biological considerations of the results, so obtained.
      - Half Window Size for Pre-filtering (in base pair)—This is the size of a half window, in which the intensities of the other probes of a local maximum will be analyzed with reference to the pre-filtering value, defined above. If no signal from the neighborhood of a probe defined by this half window is equal or higher than the pre-filtering value, the probe is considered as background noise and excluded from further consideration. We suggest using a number around half of the typical size of a DNA fragment. The default value is 200 bp.

   (b) For each candidate local maximum, the triangle model will try to fit a triangular shape with the local maximum at its peak. The range of the triangle model can vary, and the optimal range is selected on the goodness of fit by the standard deviation of the residuals from the regression model. Parameters (Fig. 1, blue arrows) relevant for this steps are as follows:

- Unit Length for Search (in base pair)—This number specifies the density of the search grid. If the unit length is defined as N, then the search range will be separated into N, 2 N, 3 N, etc. Only triangles whose size is an integer multiple of N will be considered. A small number could slightly increase the precision of the fitted triangle and the peak position, but it will increase the computational burden. The default value is 100 bp.
- Largest Search Range (in unit)—User specifies the largest number of unit length for search, defined above. For the largest search range, it is recommended that two to four times the average length of DNA fragment should be used. For an average length of ~500 base pairs and three times such length, the largest search range will be 1,500 bp or 15 × 100-bp unit. The default value is 15 units.
- Smallest Search Range (in unit)—User specifies the smallest number of unit length for search, defined above. This value should be chosen such that the average number of probes within the range specified by this number (i.e., the current probe position ± the Smallest Search Range × unit length for search) should be reasonably large for the regression analysis to have a decent degree of freedom. Default value is 3, i.e., 300 bp, which roughly includes 1 + 3 + 3 = 7 probes in the Smallest Search Range when the spacing between probes is 100 base pair.
- Range for Peak Position Perturbation (in base pair)—For a given local maximum, the program searches a small neighborhood of this local maximum and marks the site with the best triangle model fit (measured by the SD of the residuals in the regression analysis) within the neighborhood, i.e., best fitting as a peak(s). This parameter specifies the radius of this neighborhood. A number between 200 to 500 bp is recommended. The default value is 200.

(c) Post-processing of Peaks: Smaller peaks detected in the screening could be either artifacts due to a nearby large peak or false positives due to background noise. Therefore, a post-processing statistical step is performed to exclude such artifacts or false-positive peaks. The parameters

(Fig. 1, green arrows) important for this post-processing step are as follows:

- Tolerance Value of Triangle Fitting—This is used to determine the inhibition of a smaller peak by a nearby larger peak. A (smaller) peak, which is within the range of the fitted triangle of a nearby (larger) peak, is regarded as "explained by the nearby peak," if the fitted value of this probe (smaller peak) in the best fitted triangle of this probe (smaller peak) is smaller than that of the larger peak plus some amount. The amount is defined as the residual standard deviation of the fitted triangle of the nearby (larger) peak times the tolerance level, specified here. The larger/smaller tolerance value will lead to more/less number of small peaks being filtered (excluded). A value of 1–1.5 (i.e., 1–1.5 standard deviation) is generally acceptable.
- Threshold of p-value for trimming—Nonsignificant peaks with raw p-value larger than this threshold will be filtered (excluded). A threshold p-value of 0.01 is commonly accepted as a threshold for trimming.

Brief description of each of these parameters can be viewed by clicking the small button with a "?" sign on the right-hand side of each parameter entry row.

3.2 Analysis of Target Genes from the Mpeak Output Files

3.2.1 Output Files

There are 3–4 output files from the Mpeak program, depending on whether the analysis is for a single or a replicated ChIP-chip experiment. They all can be saved as GFF files and tab-delimited text files with specific names defined by the user. The files are as follows:

1. Output_shape1—Contains the chromosome coordinates and corresponding Mpeak-adjusted hybridization intensities for the entire genome. For replicated experiments, there will be a second file, Output_shape2.
2. Output_Peaks_Only—Contains the chromosome coordinates and intensities of the corresponding peaks, presented in Output shape1 and 2 files. They are usually represented by one promoter tile at the highest point of each identified peak. Occasionally, there might be two bars representing two adjacent peaks within the same promoter of a target gene.
3. Adjusted_Pvalue—Contains the raw as well as adjusted *p*-value, using the method proposed by Benjamini and Hochberg [25], of each peak identified by the program. They are positioned at the same coordinates as those displaying the corresponding peak in the Output_Peaks_Only file.

In general, the Mpeak results could be displayed with SignalMap program (Roche-NimbleGen Inc.) directly with the annotated promoter tiling GFF file of the reference genome from the respective vendors. Most of our experience is derived from ChIP-chip experiments performed with promoter microarrays of mouse MM8 and human HG18 reference genomes from Roche-NimbleGen, Inc. These promoter master lists define the chromosome coordinates of the genes, start and end points of the promoter tiles, coding strands, GeneID, gene names and official symbols, and other aliases. They are presented in Microsoft Excel and GFF formats, and can be downloaded from the Roche-NimbleGen, Inc. Web site (http://www.nimblegen.com/index.html).

3.2.2 Generation of a Consolidated File Containing the Identities, Relative Peak Binding Intensities, Statistical Significance of all Target Genes

One can take advantage of the mostly single entry-per-gene Peaks_Only GFF file and use the Microsoft Excel program to identify the target genes in reference to the corresponding master promoter-gene list from the vendor. To provide a relative relevance and statistical significance of each target gene, the corresponding hybridization intensity, hereby referred as Mpeak score, and the adjusted *p*-value can be derived from the Peak_Only and the adjusted_Pvalue file, respectively. Additional information, such as coordinates and coding strand for the structural gene, promoter tiling, and peak position, could also be deduced from the master list for each target. The assembled Excel file provides a comprehensive view of the target genes in a genome-wide manner and constitutes a dataset for additional downstream studies, including gene ontology, pathway and network analyses, and target comparisons among related transcription factors and/or different tissues, time points, and other experimental conditions.

3.3 Analysis of ChIP-Chip Signals with the Genome Browser

3.3.1 Brief Introduction to UCSC Genome Browser

The Genome Browser at the University of California Santa Cruz Genome Center portal (http://www.genome.ucsc.edu/) is an excellent online resource for analysis of results from genome-wide microarray and next-generation sequencing strategies. Numerous datasets, such as tissue-specific expression patterns, transcription factor binding, and modified histone distributions, are available, in addition to the standard gene organization, sequences, conservation, and other information on various reference genomes [12]. Importantly, custom tracks can be uploaded to Web site and analyzed together with these valuable resources. WIG files are most convenient means to generate custom tracks for Mpeak-fitted signals, such as those presented in the Output Shape GFF files.

3.3.2 Procedure to Generate WIG Files Specifying Custom Tracks for ChIP-Chip Signals

1. Open Output_Shape GFF files with Microsoft Excel.
2. Generate 4 columns, in the order of chromosome, start and end coordinates, and the hybridization intensity of each promoter tile.
3. Perform a sort on the data first on the chromosomes and then on the start of the promoter tiles in ascending order.

4. Save the data as a tab-delimited text (Tab) file.
5. Open the Tab file with a text-editing program, and paste the following at the beginning of the text file:
track type=wiggle_0 name="Name of the file" description="Description of data" useScore=1 visibility=full color=000,300,000 maxHeightPixels=128:64:0.
The name and the description of the data are inserted as underlined. The color is defined by decimal RGB color codes, and the maximum height of the display bar is defined in pixels. They can be changed according to user preferences. Save the file as a Tab file.
6. Do the same procedure with the reference genome promoter tiling, defined by chromosome locations of the start and end of the coverage on the promoters. Assign a value of 1 for all entries at the fourth column corresponding to that for the hybridization intensities in the Output_Shape file.
7. Upload the text data files to the UCSC Genome Web site as custom track files and visualize the results with the Genome Browser.

3.3.3 Discussion on Custom Track Analysis with Genome Browser

1. A full vertical viewing range for data view and inclusion of lines at zero marks for the *X*- and the *Y*-axes could provide consistent and comparable viewing among the chromosomes and different target genes on the Genome Browser.
2. For the reference file defining the promoter tiling range, **step 6** of the above procedure, the full vertical viewing range could be set at 8 or 16 pixels, with or without a line at the zero mark, according to user preference.
3. The Genome Browser offers ready access to genomic sequences of the transcription factor-binding peak regions, which can be extracted from the reference genome and analyzed in-house with sequence analysis programs, such as MacVector, for designs of primers for gene-specific PCR on binding confirmation and PCR cloning of promoter for reporter gene assays. The designed primers could be immediately evaluated with the in silico PCR analysis tools and visualized together with the ChIP-chip binding patterns in the Genome Browser. The displayed graphics can be recorded with a PDF/PS graphic download from the UCSC Genome Web site.
4. ChIP-chip strategies are primarily designed to identify target genes of transcription factors and transcriptional complexes, which frequently bind to the immediate promoter and first exon/intron regions of most genes. The potential effects of such bindings could best be correlated to the expression patterns of the target and other genes in the cells in a global manner. Genome-wide microarray expression data are widely

available in the Gene Expression Omnibus (GEO) Database at NCBI and can be downloaded for further analyses. These datasets contain global expression patterns under various conditions, such as with and without the subject transcription factor expression and differential expression between two distinct samples/cell types and/or among specific time points. They can be statistically analyzed to show differential expression ratios between two samples related to the tissues/cell types and compared with the global protein–DNA or protein complex–DNA binding patterns detected by ChIP-chip experiments.

5. Formulation of differential expression patterns could be displayed adjacent to the promoter tiling region, such as the transcription start site, of each gene. For differential gene expression ratios, such signals could be presented in a log2 format, which provides positive and negative values corresponding to differential expression favoring one or the other cell populations/tissues.

3.4 Examples of Mpeak Detection of Binding Peaks and Genome Browser Displays

For the past few years, we have used Mpeak extensively in our ChIP-chip experiments in identifying the targets for various transcription factors and in a variety of cell types and tissues of both human and mouse origins. The NimbeGen RefSeq promoter tiling microarrays for both mouse and human were used exclusively in these studies. Our results have revealed key targets and pathways associated with the biological functions of the cell types and tissues. One of these studies was designed to identify the targets for the sex-determining factor, SRY, encoded by the Y chromosome in sex determination during mouse embryonic development. Although the results will be reported in details elsewhere, some examples could be illustrated here to demonstrate the Mpeak analysis of such data, and the custom track features of the Genome Browser.

3.4.1 An Example of Peak Identification Using Mpeak and Signal Displays in Genome Browser

Total signals from the raw data on normalized and relative hybridization intensities and the Mpeak-fitted signals were tabulated as WIG files, as described above. They were then uploaded to the Genome Browser as custom tracks. Figure 2a shows the raw signals (blue bars) for promoter tiles (red) on five genes located at XqA7.3 region of the mouse X chromosome (upper two rows). Mpeak modeling identified L1cam and Avpr2 as two putative target genes (green bars) while Arhgap4, Ard1, and Renbp genes did not meet the criteria and were not selected as targets (Fig. 2a, lower three rows). Figure 2b shows the enlarged view of the promoter tiling region at L1cam locus, illustrating the initially raw signals (blue bars, top two rows), Mpeak-fitted signals (green bars, lower 2 rows), and the position of the triangle peak (dark brown bar, bottom row). Figure 2c shows the Genome Browser graphic display for one of the negative genes, Ard1, illustrating the distribution of hybridization signals (blue bars) in the raw data, which were not identified by Mpeak as a target with significant binding peak.

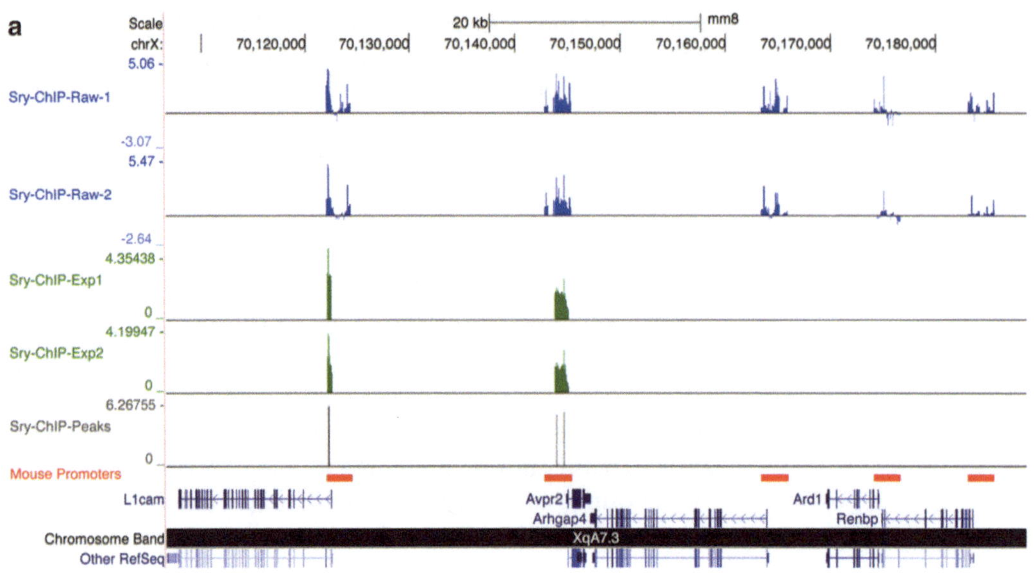

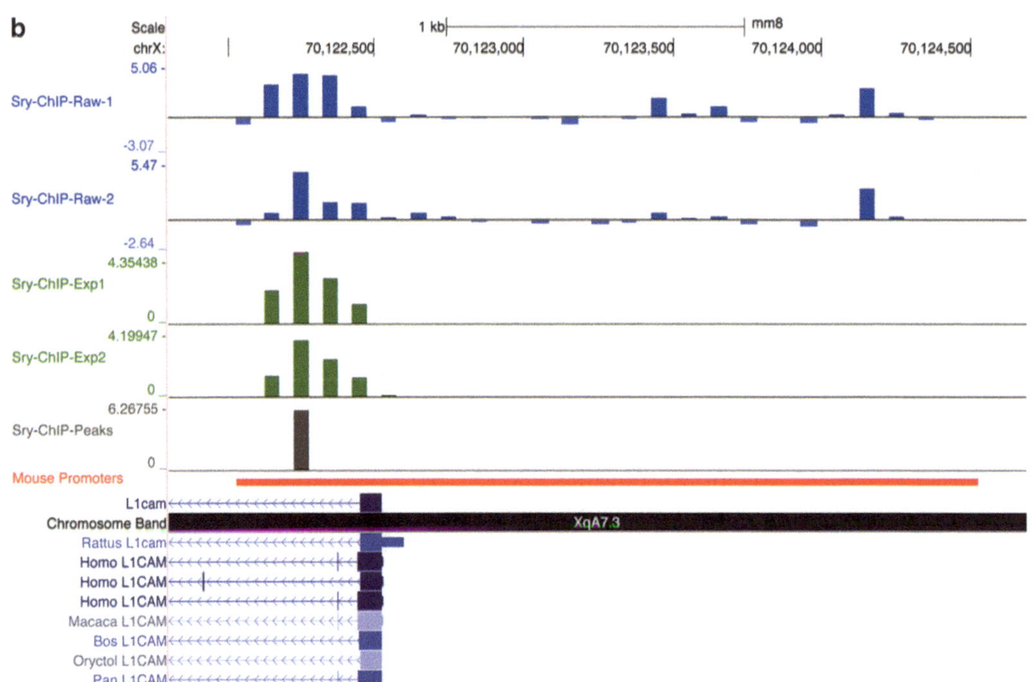

Fig. 2 (**a**) A Genome Browser graphic display of raw hybridization intensities (*blue*, *top* two rows), Mpeak-fitted binding signals (*green*, *lower* two rows), and peak (*dark brown*, *bottom* row) for five genes at XqA7.3 from an SRY ChIP-chip replicated experiment on fetal gonads of mouse embryos at the time of sex determination. Mpeak identified only two of the five genes as putative targets. (**b**, **c**) Magnified views of L1cam, a positive target, and Ard1, a negative target, from (**a**). The *red line* shows the extent of the promoter tiles for each gene. Various features, such as transcription direction, exon–intron, chromosome band, conservation of gene structures, and DNA scale, were derived from the Genome Browser

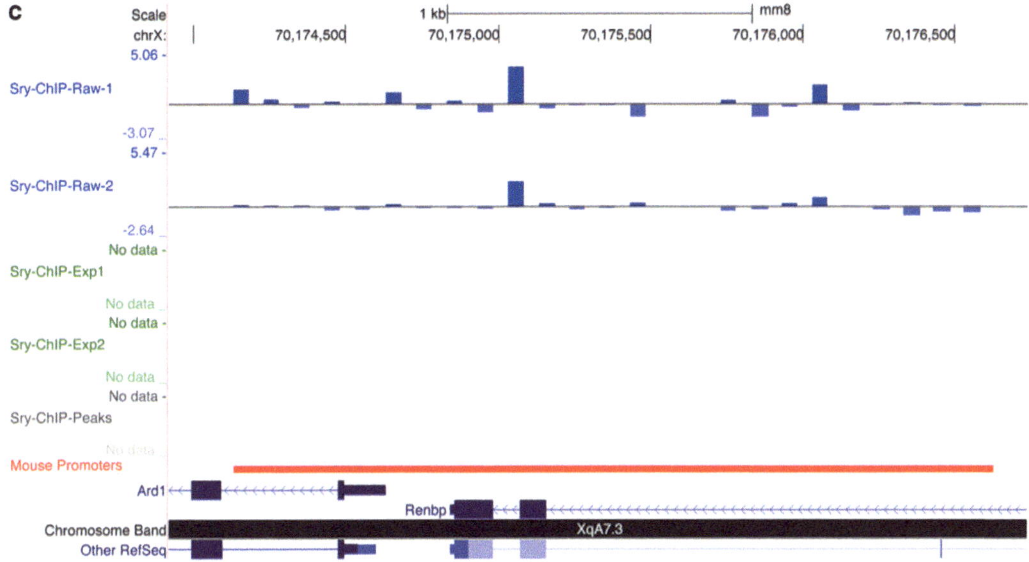

Fig. 2 (continued)

3.4.2 An Example of Comparative Analysis of Differential Gene Expression and ChIP-Chip Targets

Several groups have performed comprehensive microarray expression analyses on gonadal cells during the time of sex determination in the mouse [26–28]. Most of these datasets have been deposited in the GEO database. One of them, performed by Beverdan and Koopman [26], examined the expression patterns of somatic cells of the fetal gonads of mouse embryos at E10.5 and E11.5 stages. The probe intensity data generated from all arrays were read into the R software environment directly from the .CEL files using the R/affy package [29], which was also used to extract and manipulate probe-level data to assess data quality and to create expression summary measures. Normalization was carried out using the robust multiarray average (RMA) method [30] to generate one expression measure for each probe set on each array. The expression levels of probe sets in male gonads at 11.5 days post coitum (E11.5) were compared to those in female gonads at the same time point using t-test. The empirical Bayes method [31] was applied to adjust the standard deviation estimate for each probe on the chip. Multiple-testing correction was done using the method described [25]. We have compared the expression patterns of the male and female somatic cells at E11.5 stage (MvF, Fig. 3) and differential expression levels between male somatic cells at E11.5 and E10.5 stages (E11.5v10.5-M, Fig. 3). We have formulated the differential expression ratios between these cell types and developmental stages in log2 format, and positioned the results as 100-bp bars at the transcription start sites of the corresponding genes. Using chromosome X as an example, one can visualize the overall distribution of Mpeak-fitted peak signals along the entire length of the chromosome (Fig. 3a, lower rows, green bars). Interestingly, these signals seem to correlate with those of differential expression patterns between male and female somatic cells (MvF, Fig. 3a, reddish brown bars), and between E11.5 and E10.5 male somatic cells (E11.5v10.5-M, Fig. 3, pink bars), suggesting that SRY bound to transcriptionally active regions of the genome of somatic cells of the

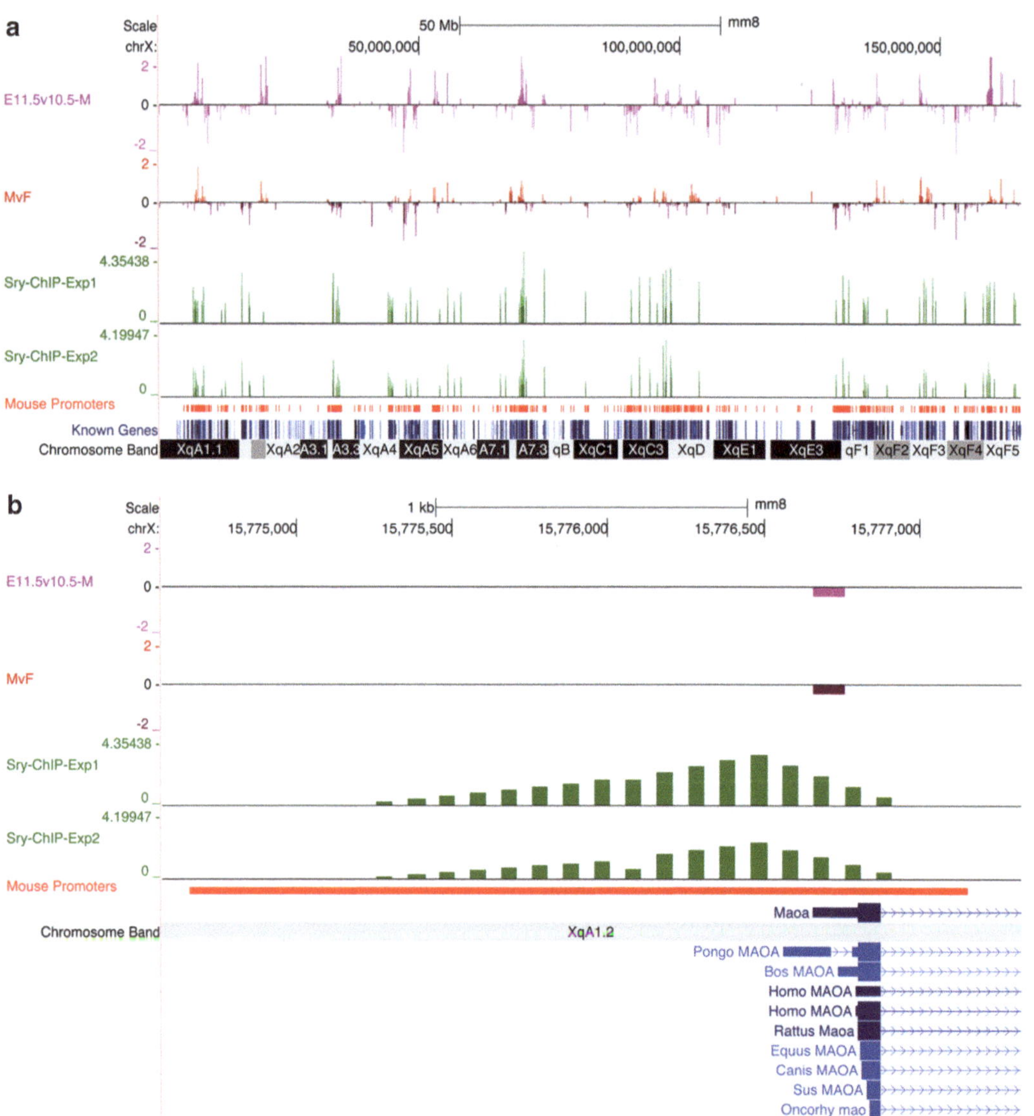

Fig. 3 (**a**) Comparison of ChIP-chip data with expression data from mouse fetal gonads at the time of sex determination, exemplified by the X chromosome and displayed with the UCSC Genome Browser. Differential expression patterns (*top* two rows) between E11.5 and E10.5 male somatic cells (E11.5v10.5-M, *pink*) and between somatic cells of male and female gonads (MvF, *reddish brown*) are displayed together with the SRY ChIP-chip Mpeak-fitted signals (*lower* rows, *green*). The SRY-binding signals are closely associated with transcriptionally active regions of the mouse genome. (**b**, **c**) Magnified views of the differential expression and SRY-ChIP-chip signals at Maoa and Fgf13 loci, respectively. Maoa is slightly down-regulated and Fgf13 is up-regulated in the male cells in the mouse fetal gonads at the time of sex determination. The differential expression patterns were derived from expression microarray analysis of Beverdam and Koopman [26]

Fig. 3 (continued)

gonads at the time of sex determination. Figure 3b, c shows the gene-specific distributions of SRY binding peaks at the promoter tiles of the monoamine oxidase (Maoa) and fibroblast growth factor 13 (Fgfl3) with down- and up-regulated transcription activities in male somatic cells, respectively. The Genome Browser offers information on the relative conservation of promoter arrangement on the genome of other species, such as human, bovine, and rat. Such conservation of promoter, transcription start site, and gene organization suggest that similar SRY-binding site(s) could be present in the respective genes of other mammalian genomes. Indeed, detailed studies of the human MAOA promoter demonstrated that the human SRY bound to similar region of the human promoter and could exert important effects on the regulation of MAOA gene by the resident SP1 transcription factor [32].

4 Notes

For researchers who are not familiar with the software used in the analysis procedure, here are some tips that may help if unexpected error happens.

1. GFF file is actually a text file that can be edited using various text-editing software. It should be noted that its format should be followed strictly to avoid unexpected reading errors in Mpeak. That is, comment lines should start with "#" and appear at the top of the file and each data row should be tab-delimited and contain exactly nine entries. For example, users will get an error message of "line 9 in file myData. GFF has bad

format" if the nineteenth line in the file is not tab-delimited or contains a number of entries other than 9. The original output file from the manufacturer's machine/software usually strictly follows such format and can be fed to Mpeak directly.

2. When replicated experiments are to be analyzed, the number of probes on each chromosome should be the same and the order that these probes appear in the files should also be identical. When the error message of "error: input replicates inconsistent: they should be with the same probes on the same chromosome" or unexpected peak detection results are obtained, please double-check the input files. Usually, if the procedure to perform replicated experiments is kept the same, such requirements are satisfied automatically.
3. Users are always encouraged to try a few different parameter settings to investigate the robustness of the software, especially "Threshold for pre-filtering" and "Threshold of *P*-value for trimming." The latter one is just the *p*-value cutoff on the resulting peaks and therefore, the larger the cutoff, more peaks are returned. The former one controls those false positives consisting of multiple aggregated probes with small signal intensity. Therefore, in general, the larger the threshold, the smaller the number of detected peaks. But it is still possible (though rare) that a peak can only be detected using higher threshold. Such peaks are usually borderline peaks whose *p*-values are not strong and may be eliminated after multiple-testing adjustment. Since more raw peaks are called under lower threshold, the penalization to the *p*-value is larger and as a result, such borderline results may be filtered.
4. Mpeak is developed under that general assumption that the spacing of the probes is around 100 bp. Although it also works under other scenarios, the user is encouraged to contact the authors to modify the software to obtain more accurate peak detection results, if needed.

Acknowledgement

This work was partially supported by a merit grant from the US Department of Veterans Affairs. YFCL is a Research Career Scientist of the DVA.

References

1. Lander ES, Linton LM, Birren B, Nusbaum C, Zody MC, Baldwin J et al (2001) Initial sequencing and analysis of the human genome. Nature 409:860–921
2. Gonzaga-Jauregui C, Lupski JR, Gibbs RA (2012) Human genome sequencing in health and disease. Annu Rev Med 63: 35–61

3. Chang HW, Chuang LY, Tsai MT, Yang CH (2012) The importance of integrating SNP and cheminformatics resources to pharmacogenomics. Curr Drug Metab 13:991–999
4. Reyes A, Semenkovich NP, Whiteson K, Rohwer F, Gordon JI (2012) Going viral: next-generation sequencing applied to phage populations in the human gut. Nat Rev Microbiol 10:607–617
5. Luciani F, Bull RA, Lloyd AR (2012) Next generation deep sequencing and vaccine design: today and tomorrow. Trends Biotechnol 30:443–452
6. Rizzi E, Lari M, Gigli E, De Bellis G, Caramelli D (2012) Ancient DNA studies: new perspectives on old samples. Genet Sel Evol 44:21
7. Shokralla S, Spall JL, Gibson JF, Hajibabaei M (2012) Next-generation sequencing technologies for environmental DNA research. Mol Ecol 21:1794–1805
8. Casals F, Idaghdour Y, Hussin J, Awadalla P (2012) Next-generation sequencing approaches for genetic mapping of complex diseases. J Neuroimmunol 248:10–22
9. Paux E, Sourdille P, Mackay I, Feuillet C (2012) Sequence-based marker development in wheat: advances and applications to breeding. Biotechnol Adv 30:1071–1088
10. Mueller TF, Mas VR (2012) Microarray applications in nephrology with special focus on transplantation. J Nephrol 25:589–602
11. Lucas SM, Heath EI (2012) Current challenges in development of differentially expressed and prognostic prostate cancer biomarkers. Prostate Cancer 2012:640968
12. Kuhn RM, Haussler D, Kent WJ (2013) The UCSC genome browser and associated tools. Brief Bioinform 14(2):144–161, Epub ahead of print
13. da Huang W, Sherman BT, Lempicki RA (2009) Systematic and integrative analysis of large gene lists using DAVID bioinformatics resources. Nat Protoc 4:44–57
14. Giardine B, Riemer C, Hardison RC, Burhans R, Elnitski L, Shah P et al (2005) Galaxy: a platform for interactive large-scale genome analysis. Genome Res 15(10):1451–1455
15. Liu T, Ortiz JA, Taing L, Meyer CA, Lee B, Zhang Y et al (2011) Cistrome: an integrative platform for transcriptional regulation studies. Genome Biol 12:R83
16. Kim TH, Ren B (2006) Genome-wide analysis of protein-DNA interactions. Annu Rev Genomics Hum Genet 7:81–102
17. Barski A, Zhao K (2009) Genomic location analysis by ChIP-Seq. J Cell Biochem 107: 11–18
18. Cuddapah S, Jothi R, Schones DE, Roh TY, Cui K, Zhao K (2009) Global analysis of the insulator binding protein CTCF in chromatin barrier regions reveals demarcation of active and repressive domains. Genome Res 19:24–32
19. Jothi R, Cuddapah S, Barski A, Cui K, Zhao K (2008) Genome-wide identification of in vivo protein-DNA binding sites from ChIP-Seq data. Nucleic Acids Res 36:5221–5231
20. Zheng M, Barrera LO, Ren B, Wu YN (2007) ChIP-chip: data, model, and analysis. Biometrics 63:787–796
21. Bolton EC, So AY, Chaivorapol C, Haqq CM, Li H, Yamamoto KR (2007) Cell- and gene-specific regulation of primary target genes by the androgen receptor. Genes Dev 21:2005–2017
22. Kim TH, Abdullaev ZK, Smith AD, Ching KA, Loukinov DI, Green RD et al (2007) Analysis of the vertebrate insulator protein CTCF-binding sites in the human genome. Cell 128:1231–1245
23. Zhang Y, Wu W, Cheng Y, King DC, Harris RS, Taylor J (2009) Primary sequence and epigenetic determinants of in vivo occupancy of genomic DNA by GATA1. Nucleic Acids Res 37:7024–7038
24. Smeenk L, van Heeringen SJ, Koeppel M, van Driel MA, Bartels SJ, Akkers RC et al (2008) Characterization of genome-wide p53-binding sites upon stress response. Nucleic Acids Res 36:3639–3654
25. Benjamini Y, Hochberg Y (1995) Controlling the false discovery rate: a practical and powerful approach to multiple testing. J R Stat Soc, Ser B 57:289–300
26. Beverdam A, Koopman P (2006) Expression profiling of purified mouse gonadal somatic cells during the critical time window of sex determination reveals novel candidate genes for human sexual dysgenesis syndromes. Hum Mol Genet 15:417–431
27. Jameson SA, Natarajan A, Cool J, DeFalco T, Maatouk DM, Mork L et al (2012) Temporal transcriptional profiling of somatic and germ cells reveals biased lineage priming of sexual fate in the fetal mouse gonad. PLoS Genet 8:e1002575
28. Nef S, Schaad O, Stallings NR, Cederroth CR, Pitetti JL, Schaer G et al (2005) Gene expression during sex determination reveals a robust female genetic program at the onset of ovarian development. Dev Biol 287(2):361–377
29. Gautier L, Moller M, Friis-Hansen L, Knudsen S (2004) Alternative mapping of probes to genes for Affymetrix chips. BMC Bioinformatics 5:111

30. Irizarry RA, Bolstad BM, Collin F, Cope LM, Hobbs B, Speed TP (2003) Summaries of Affymetrix GeneChip probe level data. Nucleic Acids Res 31:e15
31. Smyth GK (2004) Linear models and empirical bayes methods for assessing differential expression in microarray experiments. Stat Appl Genet Mol Biol 3:Article3
32. Wu JB, Chen K, Li Y, Lau YF, Shih JC (2009) Regulation of monoamine oxidase A by the SRY gene on the Y chromosome. FASEB J 23: 4029–4038

Chapter 13

Evaluation of MeDIP-Chip in the Context of Whole-Genome Bisulfite Sequencing (WGBS-Seq) in *Arabidopsis*

René Wardenaar, Haiyin Liu, Vincent Colot, Maria Colomé-Tatché, and Frank Johannes

Abstract

Studies of DNA methylation in *Arabidopsis* have rapidly advanced from the analysis of a single reference accession to investigations of large populations. The goal of emerging population studies is to detect differentially methylated regions (DMRs) at the genome-wide scale, and to relate this variation to gene expression and phenotypic diversity.

Whole-genome bisulfite sequencing (WGBS-seq) has established itself as a gold standard in DNA methylation analysis due to its high accuracy and single cytosine measurement resolution. However, scaling up the use of this technology for large population studies is currently not only cost prohibitive but also poses nontrivial bioinformatic challenges. If the end-point of the study is to detect DMRs at the level of several hundred base pairs rather than at the level of single cytosines, low-resolution array-based methods, such as MeDIP-chip, may be entirely sufficient. However, the trade-off between measurement accuracy and experimental/analytical practicality needs to be weighted carefully. To help make such experimental choices, we conducted a side-by-side comparison between the popular dual-channel MeDIP-chip Nimblegen technology and Illumina WGBS-seq in two independent *Arabidopsis* lines.

Our analysis shows that MeDIP-chip performs reasonably well in detecting DNA methylation at probe-level resolution, yielding a genome-wide combined false-positive and false-negative rate of about 0.21. However, detection can be susceptible to strong signal distortions resulting from a combination of dye bias and the CG content of effectively unmethylated genomic regions. We show that these issues can be easily bypassed by taking appropriate data preparation steps and applying suitable analysis tools.

We conclude that MeDIP-chip is a reasonable alternative to WGBS-seq in emerging *Arabidopsis* population epigenetic studies.

Key words DNA methylation, MeDIP-chip, Whole-genome bisulfite sequencing, Dye bias, *Arabidopsis*, Population epigenetics, Epigenomics

1 Introduction

DNA methylation is an epigenetic modification that involves the addition of a methyl group to the five position of the cytosine pyrimidine ring. In most animals and plants, this modification has a central role in the regulation of gene expression and in the

Tin-Lap Lee and Alfred Chun Shui Luk (eds.), *Tiling Arrays: Methods and Protocols*, Methods in Molecular Biology, vol. 1067, DOI 10.1007/978-1-62703-607-8_13,

silencing of transposable elements [1]. Because of its important biological functions, there have been substantial efforts to characterize the complete DNA methylomes of various organisms [2].

In the model plant *Arabidopsis*, the first whole-genome DNA methylation analysis used single-channel Affymetrix and dual-channel Nimblegen tiling arrays [3, 4]. These studies relied on methylation-dependent immunoprecipitation techniques followed by hybridization to high-density microarray chips (MeDIP-chip), and achieved a resolution of 35 and 220 bp, respectively. This work was instrumental in providing the first picture of the distribution of DNA methylation and its relationship with known sequence annotation in this species. A more detailed view was later obtained by several studies employing Illumina whole-genome bisulfite sequencing (WGBS-seq [5, 6]). WGBS-seq combines bisulfite conversion of DNA with next-generation sequencing (NGS) technologies and provides single cytosine resolution.

While the above-mentioned studies focused on the DNA methylome of a single reference plant, more recent work has begun to document interindividual variation in DNA methylation in large populations. The ultimate aim is to relate this type of epigenetic variation to phenotypic diversity, and to ask broader questions about the role of epigenetics in adaptive evolution. A first step in this direction was recently taken by Schmitz et al. [7] and Becker et al. [8]. These authors performed WGBS-seq on 8–12 *Arabidopsis* lines and quantified the frequency and distribution of single methylation polymorphisms (SMPs) as well as differentially methylated regions (DMRs). These experiments generated roughly 200–500 GB of data and required construction of extensive data pipelines. Scaling up the use of WGBS-seq to even larger samples poses nontrivial bioinformatic challenges that range from data storage to downstream computation analysis. These challenges can hinder the routine application of this technology for future population epigenetic studies.

A viable alternative is to restrict DNA methylation analysis to the detection of DMRs, which typically range between 10 and 1,000 bp in length. In *Arabidopsis* as in other species DMRs appear to be functionally more important than SMPs [7–10] and appear to be a suitable unit of analysis. Focusing on DMRs has the important advantage that array-based measurement technologies, such as MeDIP-chip, could be employed in place of WGBS-seq because they provide sufficient resolution. The use of array-based methods can substantially reduce the bioinformatic resources required to perform population epigenetic studies. Nonetheless, loss of measurement accuracy resulting from hybridization compared to sequencing may present a significant drawback which not all researchers are willing to accept. Furthermore, it should be noted that unlike WGBS-seq, MeDIP-chip does not allow distinguishing between CG, CHG, and CHH methylation, a point which may be

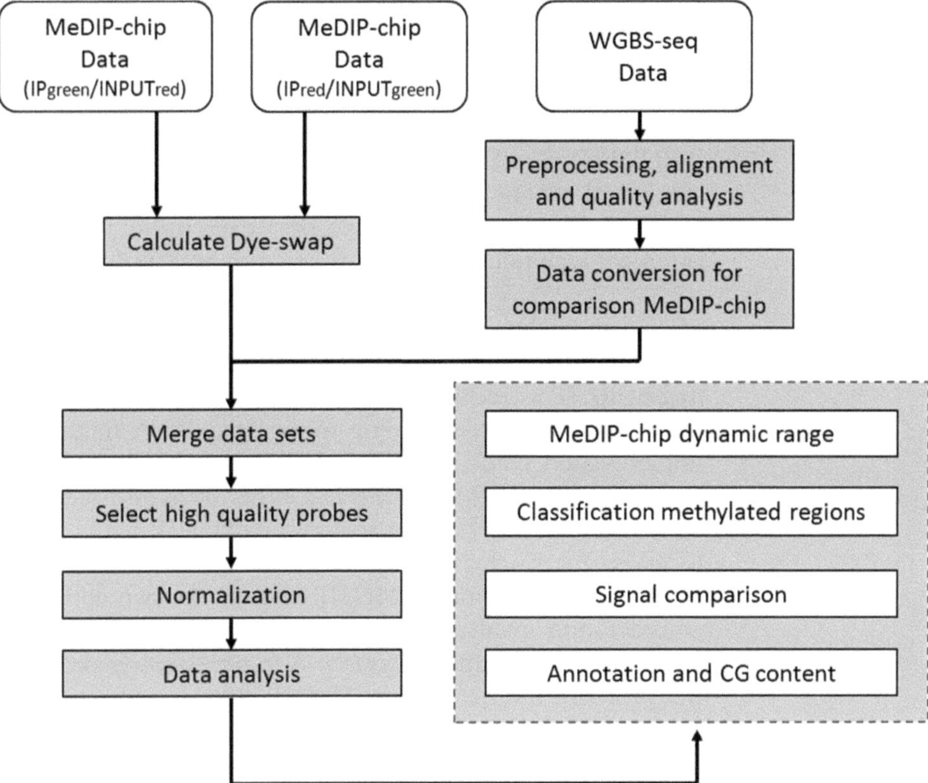

Fig. 1 Workflow evaluation MeDIP-chip

important to consider in some instances. Hence, the trade-off between loss of measurement accuracy and experimental/analytical practicality needs to be weighted carefully. To help make such experimental choices we conducted a side-by-side comparison between the popular dual-channel MeDIP-chip Nimblegen technology and Illumina WGBS-seq. The workflow shown in Fig. 1 serves as an outline of this chapter.

Our analysis shows that the dual-channel MeDIP-chip technology performs reasonably well in detection of probe-level DNA methylation, which is approximately the minimum resolution required for DMR detection. We estimate that MeDIP-chip yields a combined false-positive and false-negative rate of 0.21 genome-wide. However, we also find that detection can be critically dependent on prior data preparation steps, signal distortions arising from dye biases, and the statistical method used for detection. Based on our results, we make several simple but important recommendations regarding the experimental implementation of MeDIP-chip for population epigenetic studies (*see* **Note 1–3**).

2 Data Sets and Data Preparation

In this chapter we consider DNA methylation data from two different epigenetic recombinant inbred lines (epiRILs; R60 and R202 [11]). The DNA methylomes of each epiRIL were measured using dual-channel Nimblegen MeDIP-chip and Illumina WGBS-seq [12]. This section provides a brief overview of these two measurement technologies as well as key data preparation steps.

2.1 MeDIP-Chip

Methylated DNA immunoprecipitation (MeDIP) is a large-scale purification technique used for the enrichment of methylated DNA fragments. This technique uses antibodies specific to methylated cytosines in order to separate methylated DNA fragments from unmethylated DNA fragments. Following this separation procedure the methylated fragments can either be hybridized to a tilling array (MeDIP-chip [13]) or sequenced (MeDIP-seq [14]) in order to assess the methylation status of the genome under consideration. The application of MeDIP-chip to the two epiRILs under consideration involved several experimental steps which are discussed in short in this paragraph. A more extensive description of the protocol used to obtain MeDIP-chip data described in this chapter is given by Cortijo et al. [15].

1. *Extraction and fragmentation*—DNA was extracted from aerial parts of 3-week-old *Arabidopsis* plants using a standard extraction kit (Qiagen DNeasy plant Maxi kit). Extracted DNA was fragmented using sonication. Sonication produced fragments with a size between 100 and 600 bp (verified with gel electrophoresis).
2. *Immunoprecipitation and amplification*—After this fragmentation step, the DNA was denatured and anti-5mC antibody was added to the IP DNA pool, which recognizes specifically methylated cytosines in single-stranded DNA. Magnetic beads containing binding sites for this antibody were then added to pull down (i.e., immunoprecipitate) methylated fragments. Following release of the antibody, both IP and input DNA fractions were amplified by PCR.
3. *Labeling and hybridization*—The IP and input fractions were differentially labeled with two fluorescent dyes (Cy3 and Cy5) and hybridized to Nimblegen whole-genome *Arabidopsis* tiling arrays (3×720 K array) containing 711,320 isothermal probes. Depending on the CG content, these probes range from 50 to 75 nucleotides in length and have an inter-probe spacing of about 110 base pairs on average.
4. *Scanning the tiling array*—After hybridization, the intensities of both dyes were obtained by scanning the tiling array. The scanner outputs two files with the raw intensities for each

probe on the tiling array: one file with the IP signals and the other file with the input signals.

5. *Signal calculation*—The IP and input signals were log transformed and subtracted from each other ($\log_2(\text{IP}) - \log_2(\text{input})$). Hence, probes with a low signal are from genomic regions that show low methylation levels, and those with a high signal are from genomic regions that show high methylation levels.

2.2 Whole-Genome Bisulfite Sequencing

WGBS-seq is a NGS technology used to determine the DNA methylation status of single cytosines. In the case of WGBS-seq, unlike other NGS technologies, the DNA is treated with sodium bisulfite before sequencing. Sodium bisulfite is a chemical compound that converts unmethylated cytosines into uracil [16, 17]. Knowing which cytosines have converted it is possible to determine which cytosines are methylated (not converted) and which ones are unmethylated (converted into U). After sequencing, the unmethylated cytosines appear as thymines. There are several ways of producing sequence libraries for WGBS-seq (*see* Ref. 18). The WGBS-seq data in this chapter was produced with a technique developed by Cokus et al. [6]. Here we describe in short the procedure.

1. *Extraction and fragmentation*—DNA was extracted from aerial parts of 3-week-old *Arabidopsis* plants using standard extraction kit (Qiagen DNeasy plant Maxi kit). Extracted DNA was fragmented by sonication.
2. *Adapter ligation and size selection*—A set of double-stranded adapter sequences was ligated to the fragmented DNA. These adapter sequences contained methylated adenine bases with DpnI restriction sites. The restriction sites are important for the removal of the first set of adapter sequences in one of the subsequent steps. Gel electrophoresis was used to obtain adapter-ligated fragments with an appropriate size.
3. *Bisulfite conversion and amplification*—After the addition of the adapter sequences the sodium bisulfite conversion is performed. During this step, unmethylated cytosines are changed into uracil. PCR was subsequently performed with the use of primers that were complementary to the converted adapter sequences.
4. *Removal of first adapter sequences and ligation sequencing adapters*—After the first PCR amplification step the first set of adapters was removed using DpnI restriction enzymes. A new set of sequencing adapters was subsequently ligated to BS-converted DNA fragments.
5. *Size selection and amplification*—Fragments with a size between 120 and 170 bp were selected with the use of gel electrophoresis, and a second and final PCR step was performed using

primers complementary to the sequencing adapters to yield a sequencing library.

6. *Sequencing*—Illumina sequencing technology (Illumina 1G Genome Analyzer) was used to produce read sequences with a length of 76 or 101 nt.

   After sequencing, the reads need to be mapped (or aligned) to a reference genome in order to infer the methylation status of the cytosines of the genome under consideration. However, mapping these converted sequences is not straightforward since unmethylated cytosines will result in mismatches with the reference genome (i.e., they appear as thymines). To circumvent this issue several programs have been developed that first convert the reference genome into a three-letter genome (i.e., all cytosines are changed into thymines; in silico [19]). The remaining cytosines of the read sequences also need to be changed into thymines before mapping. The in silico-treated read sequences are subsequently mapped to this three-letter reference genome. Once the mapping has been performed, methylation status can be inferred using the original sequence of the reference genome and the read. A thymine (read) mapped to a cytosine (reference) is an unmethylated cytosine. A cytosine mapped to a cytosine is a methylated cytosine. We utilized BS Seeker [20] for the mapping of the read sequences. BS Seeker is a python-based open-source mapping program for the alignment of bisulfite-treated sequences. The analysis includes preprocessing of the reads prior to alignment, the alignment itself, and quality analysis of the data. The steps are described further below. More details of the mapping of the reads can be found elsewhere [12].

7. *Removal of adapter parts*—When a DNA fragment is shorter than the read sequence a part of the adapter sequence will also be sequenced. The adapter sequence was added artificially and does not match with the reference genome. We therefore removed this part using a sliding window approach. The part that overlapped with the known adapter sequence was removed.

8. *Removal of short reads*—The removal of the adapter sequences resulted in some cases in reads with a length smaller than 30 nucleotides. These short reads are more difficult to map and were therefore removed. The proportion of short reads was in our case quite small and therefore the final read coverage was barely affected.

9. *Removal of duplicated reads*—Duplicated reads were removed in the final preprocessing step because they were likely produced during PCR amplification and were therefore not informative.

10. *Alignment to reference genome*—After these preprocessing steps the reads were mapped to a reference genome (TAIR 10)

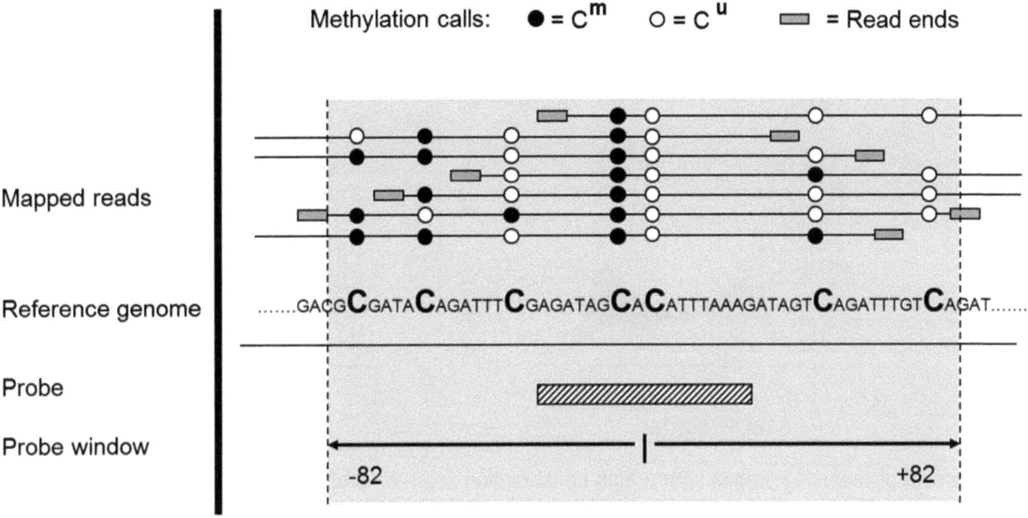

Fig. 2 Calculation of whole-genome bisulfite sequencing signals. The methylation calls within the window (*gray*) are used to calculate a normalized whole-genome bisulfite sequencing signal (see formula)

with the use of BS Seeker. After mapping the obtained average genome coverage was 29 and 27× for epiRIL 60 and 202, respectively (both strands combined).

11. *Determination conversion rate*—One important step in the quality analysis of the data is to determine the (bisulfite) conversion rate. The conversion rate, which is the percentage of unmethylated cytosines that effectively changed into uracil, was determined for both epiRILs after mapping. The conversion rates were determined with the information of reads that were mapped to chloroplast DNA. The chloroplast DNA is known to be unmethylated and therefore any detected methylated cytosine is considered to be a non-converted unmethylated cytosine. Both epiRILs showed a conversion rate above 99 % which indicates that the data is of good quality.

2.3 Data Conversion and Normalization

One of the major differences between MeDIP-chip and WGBS-seq is mapping resolution. WGBS-seq can interrogate the methylation status of individual cytosines while MeDIP-chip achieves a resolution of about 165 bps. In order to facilitate a meaningful comparison between the two technologies, we converted the WGBS-seq data into a format that is comparable to that of MeDIP-chip. To achieve this we calculated the proportion of methylation calls in windows of 165 bps centered at the probe sequence (Fig. 2). By methylation calls we mean the individual methylation calls of each read sequence. This results in a signal ranging from zero to one. This signal is afterwards normalized for the number of cytosines in the probe window. Let C_j^m denote the number of cytosines that have been called methylated in the jth window, C_j^u the number of

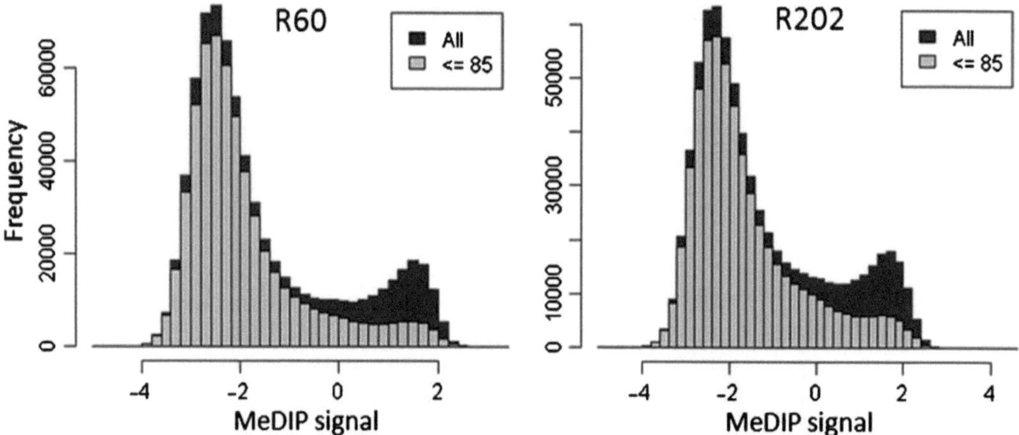

Fig. 3 Impact of removing probes with a high conservation score on probe signal distribution. Probes with a high conservation score (*dark gray*) typically show signal intensities similar to probes that correspond to genomic regions with high methylation levels

cytosines that have been called unmethylated in the *j*th window, R_j the total number of cytosines in the *j*th window according to the reference genome, and R_{max} the maximum number of cytosines across all windows. The converted and normalized WGBS-seq signal can be calculated as

$$\text{WGBS}_{\text{sig}}(j) = \frac{C_j^m}{C_j^m + C_j^u} \times \frac{R_j}{R_{\text{max}}}.$$

We selected high-quality data for the analysis described in this chapter. In case of the WGBS-seq probe windows, we only selected windows with 35 or more cytosines, with at least half of the cytosines being covered by one or more read sequences. In case of the MeDIP-chip data we only selected probes with a conservation score smaller or equal to 85. The conservation score of a probe indicates the uniqueness of a probe sequence. These scores were obtained by performing a blast search. Scores are percentage of identity with the second best hit (score range 45–100). The best hit is with the genomic location for which the probe was designed. Probes with a high conservation score are more likely to cause cross-hybridization problems. As shown in Fig. 3, removal of probes with a high conservation score has a drastic impact on the MeDIP signal distribution, as they typically show signal intensities similar to probes that correspond to genomic regions with high methylation levels. *See* **Note 1** for recommendations concerning the quality of the data.

Analysis was performed on probes (i.e., probe windows) that were present in both data sets (MeDIP-chip and WGBS-seq signal

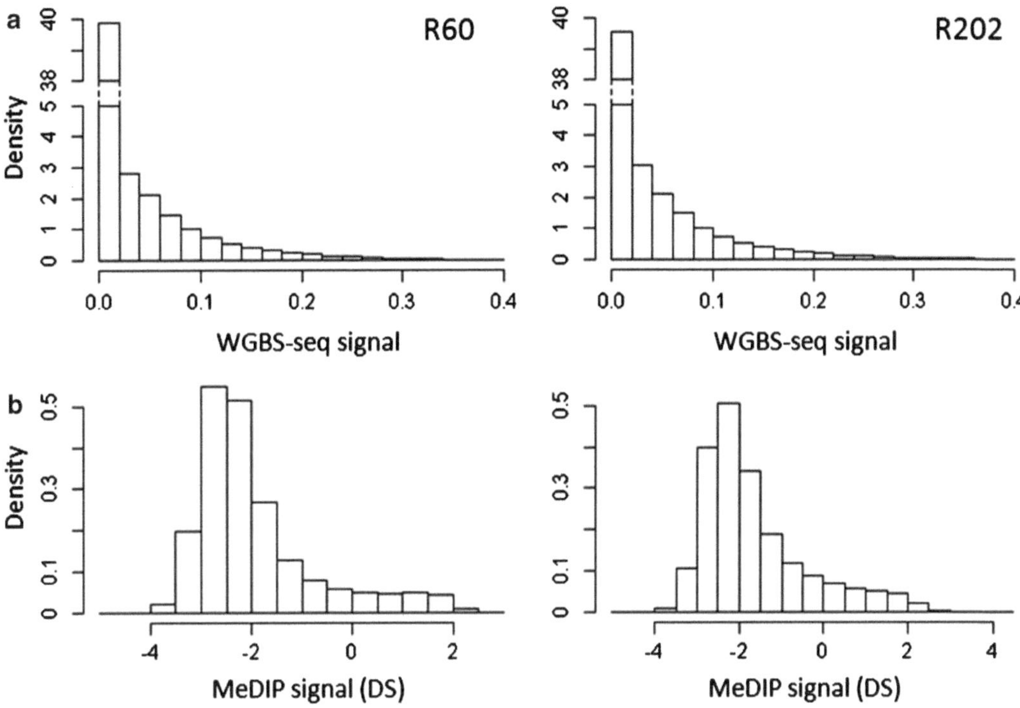

Fig. 4 Density histogram of WGBS-seq signals and MeDIP-chip signals following data conversion and normalization. (**a**) WGBS-seq signal distribution. (**b**) MeDIP-chip signal distribution

data). This yielded 551,688 and 550,676 high-quality probe windows in total, covering approximately 77.6 and 77.4 % of the genomes of R60 and R202, respectively. In case of the MeDIP-chip data a total of two dye-swap experiments were performed for each epiRIL. The log-transformed signals were also averaged over the two dye-swap experiments which resulted into three MeDIP-chip data sets:

- G/R data: IP labeled green (Cy3, G) and input labeled red (Cy5, R).
- R/G data: IP labeled red (Cy5, R) and input labeled green (Cy3, G).
- DS data: Average of G/R and R/G data (dye-swap data).

Finally, quantile normalization was applied to bring these three data sets to a common scale [21]. Figure 4 displays a density histogram of the WGBS-seq signal for the two epiRIL experiments (R60 and R202) and the MeDIP-chip signal (DS).

2.4 Software

The analysis described in this chapter was performed using R [22]. R is a command-line software environment for statistical computing and graphics. The Hidden Markov Model (HMM) for probe classification was programmed in C++ [15].

3 Results

3.1 Assessment of MeDIP-Chip Dynamic Range

Since the WGBS-seq signal provides a measure of the proportion of methylated cytosines in a given probe window, we were able to assess the dynamic range of the MeDIP-chip technology directly by empirical comparison. To achieve this we calculated the median MeDIP-chip signal for sliding windows along the entire WGBS-seq signal range (Window size: 0.05; step size: 0.01; Fig. 5a). We find that MeDIP-chip exhibits good sensitivity for low-to-intermediate methylation levels (WGBS-seq range: 0.00 to ~0.13). In this low-to-intermediate range, there is a nearly linear relationship between the WGBS-seq signal and the MeDIP median signal, but the MeDIP-chip sensitivity falls off quickly and saturates at a WGBS-seq signal value of about 0.28. Above this point, MeDIP-chip is effectively unable to differentiate between methylation levels. However, in R60 and R202 this saturation effect affects only a relatively small number of probe windows, 0.19 % (N = 1,056) and 0.22 % (N = 1,220) of all regions genome-wide, and 0.58 and 0.67 % of all methylated regions (*see* Table 1), respectively. Signal saturation should therefore not be a matter of great concern in the analysis of the *Arabidopsis* methylome. Similar conclusions can be reached when considering the MeDIP signal on its original scale (data not shown), rather than on the log-transformed scale, which indicates that saturation is not caused by scaling issues.

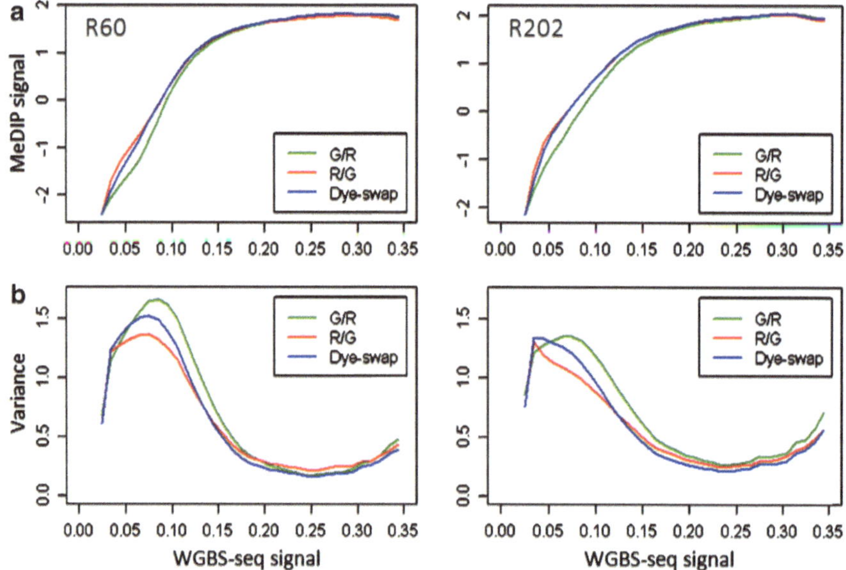

Fig. 5 Median and variance of the MeDIP signal along the entire WGBS-seq signal range. Median MeDIP signal (**a**) and variance MeDIP signal (**b**) for sliding window along the entire WGBS-seq signal range. The G/R data show less sensitivity for low-to-intermediate WGBS-seq signals and also show a higher variance compared to the R/G data

Table 1
Classification probe windows using WGBS-seq

	R60	R202
Classification cutoff	5.09E−3	7.03E−3
# of probe windows selected for analysis	551,688 (77.6)	550,676 (77.4)
# of unmethylated probe windows	369,358 (67.0)	367,526 (66.7)
# of methylated probe windows	182,330 (33.0)	183,150 (33.3)
Median signal, unmethylated windows	9.58E−4 (0.000–5.09E−3)	9.65E−4 (0.000–7.02E−3)
Median signal, methylated windows	3.32E−2 (5.09E−3–0.552)	3.33E−2 (7.03E−3–0.562)

Reported are the number of unmethylated and methylated windows using a genome-wide false-positive rate of 0.01

Our analysis also indicates that there are clear dye-related differences in the MeDIP median signal. The R/G data appears to respond more sensitively to changes in WGBS-seq methylation levels compared with the G/R and the DS data (Fig. 5a). While these dye differences disappear at the saturation point (WGBS-seq signal ~ 0.28), they are most prominent in the optimal dynamic range. This divergence is even more severe when we consider the MeDIP signal variance across the complete WGBS-seq range using the same sliding window approach (Fig. 5b). Ideally, the signal variance should be low and constant across methylation levels. Figure 5b illustrates that this is clearly not the case: the MeDIP signal variance is largest within the optimal dynamic range but decreases rapidly with increasing methylation levels. Notably, the G/R data displays a 1.20 (R60)- and a 1.28 (R202)-fold increase in signal variance (on average) relative to the R/G data suggesting that it is substantially noisier, and the dye-swap (DS) fails to correct this bias. This latter observation is contrary to what is typically seen in expression microarrays where dye-swaps have proved to be an effective strategy [23, 24]. *See* **Note 2** for recommendations concerning the labeling of the IP and input DNA.

3.2 Classification of Methylated Regions Using MeDIP-Chip

From a data analysis standpoint, the classification of individual probes according to their underlying methylation status (i.e., methylated or unmethylated) is critically dependent on the variance of the MeDIP signal. When signal variation is large, classification tends to be more difficult. Many statistical analysis methods have been proposed to minimize this problem and to facilitate accurate probe classification in the context of MeDIP-chip data (e.g., [15, 25–29]). The development of such methods continues to be an active area of research. Classification is particularly problematic in regions of the MeDIP signal distribution where signals

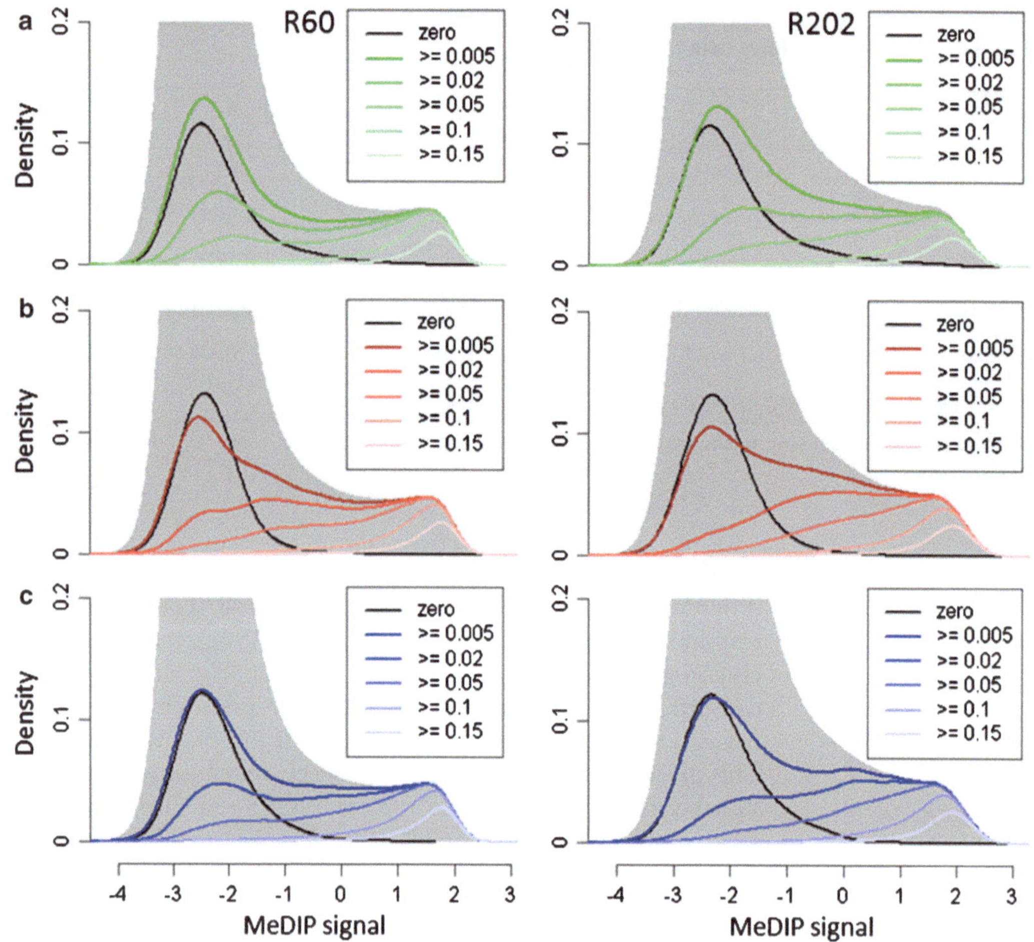

Fig. 6 Overlap unmethylated and methylated probe signal distributions. Shown are the MeDIP distributions that correspond to a certain WGBS-seq signal range. It becomes clear that there is a significant overlap between the MeDIP signal distributions with a low WGBS-seq signal (unmethylated; *black* is WGBS-seq signal of zero) and those with a high WGBS-seq signal (methylated). The G/R data (**a**) tends to have a higher overlap compared to the R/G data (**b**). The DS data (dye-swap; **c**) shows an average overlap result. It also shows a higher overlap compared to the R/G data

from unmethylated probes overlap with those from methylated probes (Fig. 6). In this case, the task is to find an informative cut-off that would minimize both false-positive and false-negative methylation calls.

3.2.1 Defining the "Gold Standard"

To illustrate this problem empirically we use the WGBS-seq signal to define high-confidence methylated probe windows based on a measurement error distribution. To achieve this, let the measurement error, y, of the jth probe window be given by

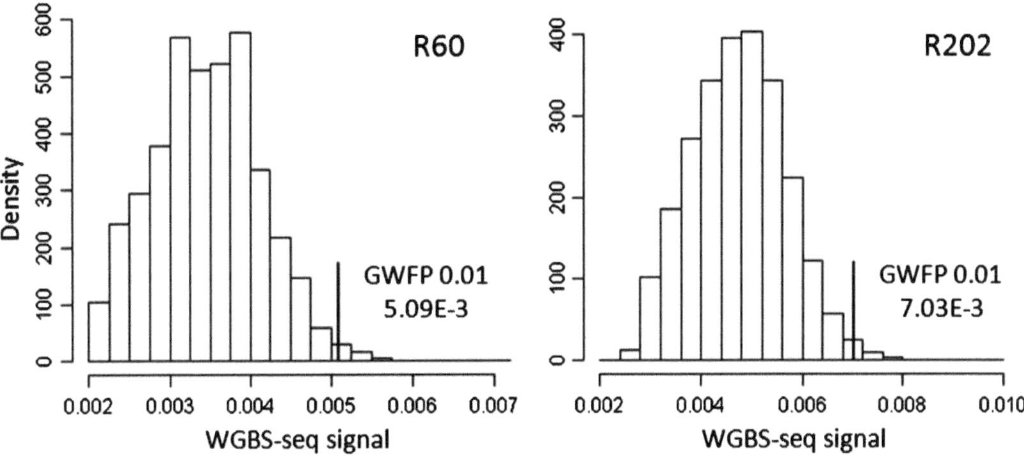

Fig. 7 Measurement error distributions. Shown are the measurement error distributions of both epiRILs and the signal cutoffs that correspond to a genome-wide false-positive rate of 0.01

$$y(j) = (1 - \mathrm{CR}) \times \frac{R_j}{R_{\max}},$$

where R_j is the number of cytosines in the *j*th probe window, R_{max} the maximum number of cytosines across all windows, and CR is the overall conversion rate. Furthermore, let us define the empirical density distribution of the error as $f(y)$ (Fig. 7). Using this distribution, the genome-wide false-positive (GWFP) rate can be calculated numerically using

$$\mathrm{GWFP} = 1 - \int_0^T f(y)\mathrm{d}y,$$

where T is the WGBS-seq signal threshold that is needed to meet a given GWFP level. We find that for GWFP = 0.01, the WGBS-seq threshold is approximately 5.09E−3 and 7.03E−3 for R60 and R202, respectively. Hence, we define probe window j as methylated if the WGBS-seq signal of that region is larger than the threshold T. At this threshold level, we find that 33.0 % (N= 182,330) and 33.3 % (N= 183,150) of all probe windows (genome-wide) can be confidently called methylated with this technology in R60 and R220, respectively (Table 1).

3.2.2 MeDIP Signal Classification Based on a Naïve Classifier

We use the WGBS-seq-derived classification to assess the problem of determining the methylation status of probes in the context of MeDIP-chip data. We first consider a naïve classifier which consists of a single MeDIP cutoff. According to this classifier, a probe is considered methylated if its signal exceeds the cutoff and as unmethylated if its signal falls below it. Comparing the

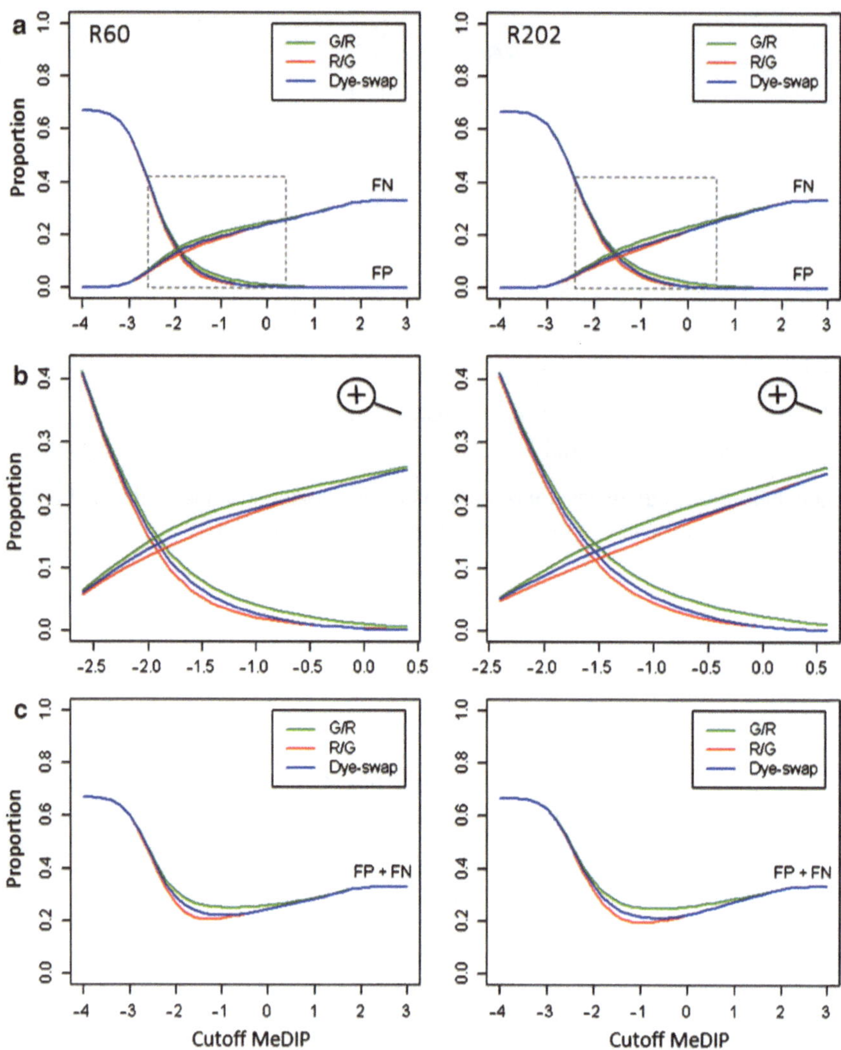

Fig. 8 False-positive and false-negative rates using a naïve classifier. Shown are the proportions of false-positive and false-negative probe classifications for different classification cutoffs of the MeDIP data (**a**) and (**b**) and the sum of the two (**c**). The G/R data shows a substantial higher proportion of FP and FN compared to the R/G data

resulting calls to those obtained from the WGBS-seq classification (Subheading 3.2.1) allows us to define the MeDIP false-positive and false-negative rates associated with the naïve classifier. Figure 8a, b shows the distribution of false-positive and false-negative rates for series of cutoffs across the entire MeDIP signal range (−4 till 3 with step size 0.2). This analysis shows that there is a considerable trade-off between minimizing false positives and false negatives while maximizing the total number of regions detected as methylated. We find that the optimal cutoff corresponds to an MeDIP signal of about−1.5 (Fig. 8), which yields a

Table 2
Number of false-positive and false-negative probe classification

	R60			R202		
	FP	FN	FP+FN	FP	FN	FP+FN
G/R—naïve cutoff	17,973	120,335	138,308 (0.251)	25,945	111,320	137,265 (0.249)
G/R—HMM	12,796	114,197	126,993 (0.230)	12,385	110,766	123,151 (0.224)
R/G—naïve cutoff	16,763	97,826	114,589 (0.208)	24,932	83,325	108,257 (0.197)
R/G—HMM	11,564	108,238	119,802 (0.217)	13,985	101,413	115,398 (0.210)
DS—naïve cutoff	14,901	107,635	122,536 (0.222)	15,852	101,227	117,079 (0.213)
DS—HMM	12,905	107,769	120,674 (0.219)	15,179	100,496	115,675 (0.210)

Reported are the number of false-positive and false-negative probe classifications using either a naïve cutoff or a Hidden Markov Model. Numbers of the naïve cutoff are based on the most optimal cutoff in Fig. 8c (smallest number of FP+FN)

combined false-positive and false-negative rate of about 0.23. Consistent with the dye-effects illustrated in our assessment of the dynamic range (Subheading 3.1), the combined false-positive and false-negative rates show substantial dye-dependence, with genome-wide rates of about 0.25, 0.20, and 0.22 for the G/R, R/G, and the DS data, respectively (Fig. 8; Table 2). Hence, the G/R data yields the highest misclassification rate which is likely caused by its high signal variance. Again, the dye-swap does not correct this problem (Fig. 8). *See* **Note 2** for recommendations concerning the labeling of the IP and input DNA.

3.2.3 MeDIP Signal Classification Based on Hidden Markov Model

This dye bias can be partially alleviated if one considers, instead of the above naïve classifier, a more sophisticated statistical classification approach based on HMMs (Fig. 9, Table 2). We have recently proposed an HMM for the analysis of MeDIP-chip [15]. This model has been shown to outperform alternative methods in terms of speed, sensitivity, and specificity [28]. An important aspect of this model is, as with HMMs in general, that it borrows signal information from immediately surrounding probes, and therefore significantly reduces measurement noise. This leads to a more robust inference of the underlying methylation status of a given probe window and makes methylation analysis less susceptible to dye effects. Figure 9 illustrates this point clearly; it shows that the HMM analysis of the R/G, G/R, and DS data results in much smaller misclassification differences between these data sets in terms of overall false-positive and false-negative rates (about 0.23, 0.21, and 0.21 for the G/R, R/G, and the DS data, respectively; Table 2). *See* **Note 3** for recommendations concerning the

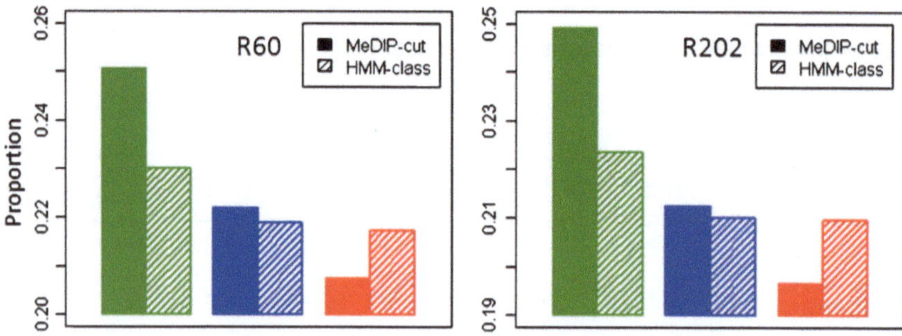

Fig. 9 Comparison performance of a naïve classifier and a Hidden Markov Model. Shown are the proportion of misclassified probes (FP + FN) obtained using the most optimal MeDIP classification cutoff (MeDIP-cut; smallest number in Fig. 8c) or the HMM classification (HMM-class). The color of each bar corresponds to the three data sets (*green*: G/R data; *red*: R/G data; *blue*: DS data)

analysis of the data. Nonetheless, despite this improvement, dye-related differences, particularly in the G/R data, do persist and continue to affect our ability to infer the correct methylation status of a given genomic region (Fig. 9). It is therefore of interest to identify and characterize the sources of this bias in the MeDIP-chip data.

3.3 Dye Bias in MeDIP-Chip Is Associated with Low Methylation Levels and CG Content

To explore the source of the observed dye bias, we start by plotting the two dye combinations (G/R and G/R) in a scatter plot (Fig. 10a). We find that there is a subset of the probes that shows a relatively higher signal for the G/R data compared to the R/G data. Inspection of the WGBS-seq signal corresponding to these probes indicates that the methylation level of these probe windows should be low (high density of probes around zero). This expectation is indeed reflected in the R/G signal (Fig. 10c), but not in the G/R signal which seems to be vastly exaggerated (Fig. 10b). Since the dye-swap signal yields only an average of the R/G and G/R data it cannot correct this bias (Fig. 10d).

Annotation analysis of this subset of probes shows that they contain a high proportion of transposon and intergenic sequences relative to genic sequences (Fig. 11a). In *Arabidopsis*, it is well known that genes have a higher CG percentage compared to transposons (Fig. 12). This raises the question whether CG content may be a key contributor to the observed dye bias.

In order to explore this possibility more generally, we calculated the CG content of the probe window for each probe on the tiling array and examined its relationship with signal intensity in the G/R, R/G, and DS data sets. For clarity we restricted our analysis to probes that were unmethylated according to WGBS-seq (*see* Table 1). In this way we could rule out any trends arising from differences in methylation levels. Our analysis shows that the signal intensity of unmethylated regions in the G/R data is subject

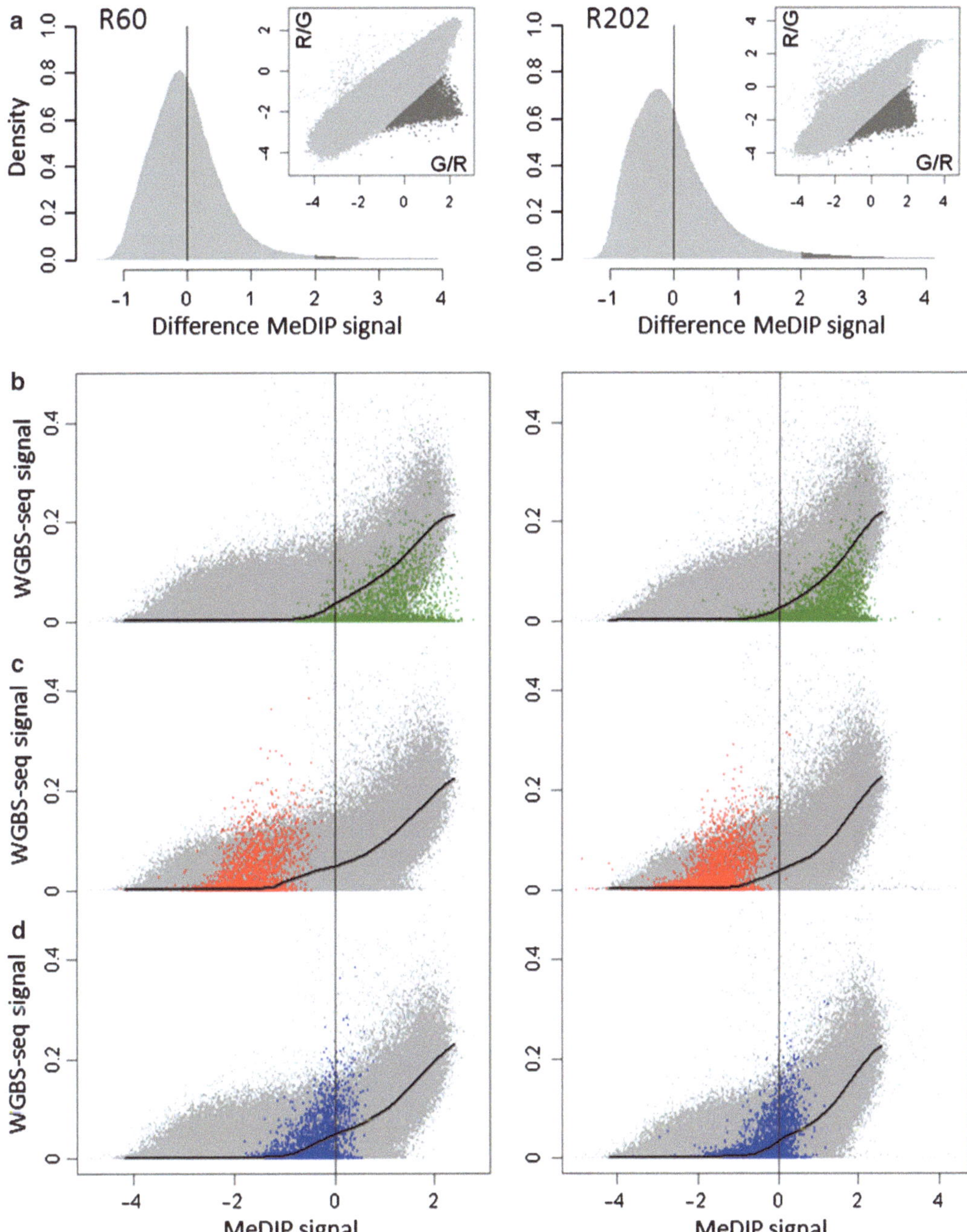

Fig. 10 Inspection of non-correlating probes using WGBS-seq signal data. (**a**) Shown is a density plot of the difference of both dye combinations. The inset shows a scatter plot of both dye combinations. Probes with a signal difference higher than two are indicated with dark gray. (**b**)–(**d**) Scatter plots of WGBS-seq data (*y*-axis) and each of the three MeDIP data sets (*x*-axis). The non-correlating probes (*dark gray* in panel **a**) are highlighted according to the color assigned to each data set (*green*: G/R data; *red*: R/G data; *blue*: DS data). The *black line* shows the median WGBS-seq signal for sliding windows along the entire MeDIP signal range

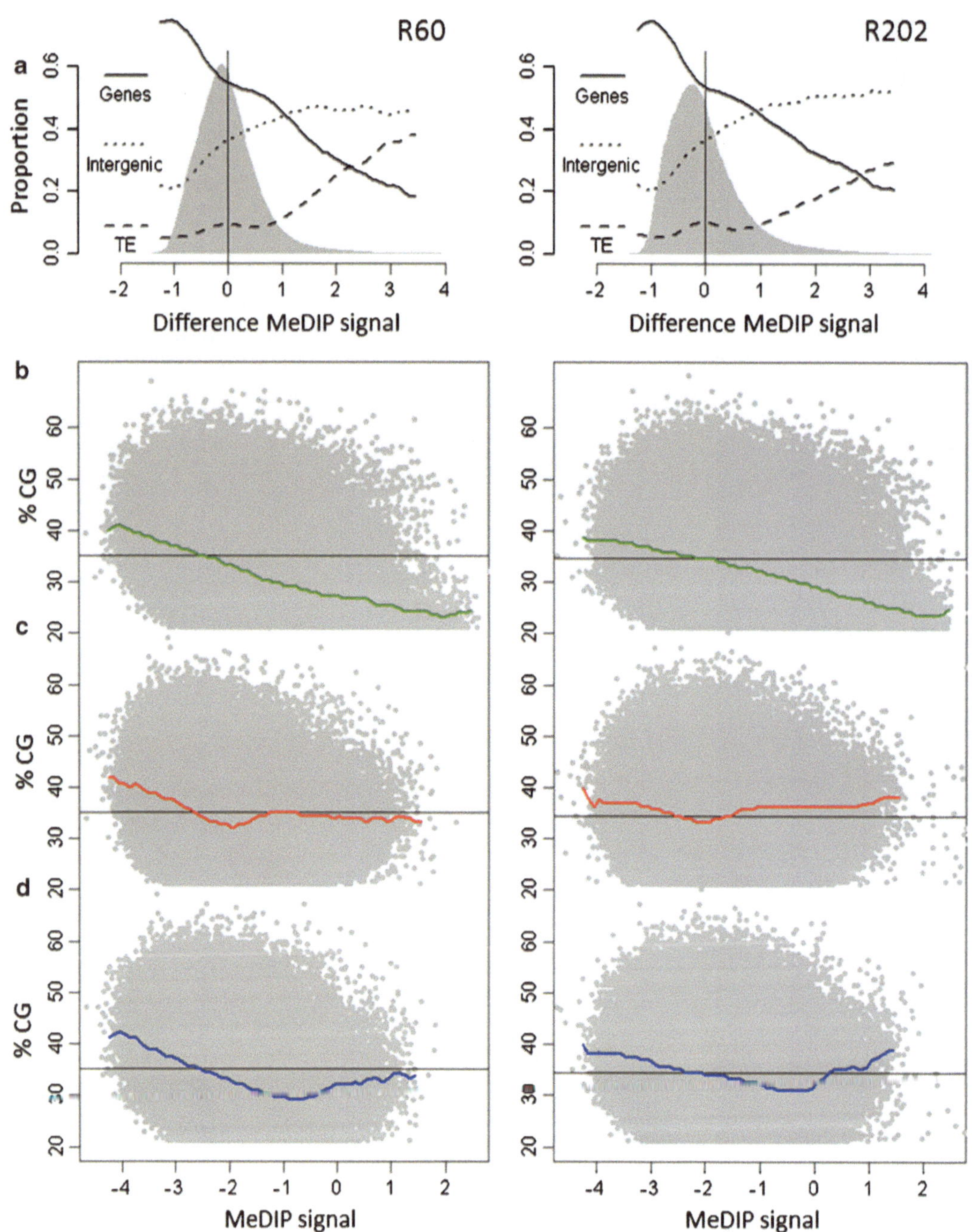

Fig. 11 Annotation analysis and CG bias of unmethylated probes. (**a**) Shown is the proportion of genic, transposable element (TE) and intergenic probes along the entire MeDIP difference range (G/R–R/G). (**b**)–(**d**) Scatter plots of the CG percentage of each probe window (*y*-axis) and the MeDIP signal of each of the three MeDIP data sets (*x*-axis). The *trend lines* show a clear negative relationship between CG content and signal intensity. The color of each line corresponds to the three data sets (*green*: G/R data; *red*: R/G data; *blue*: DS data)

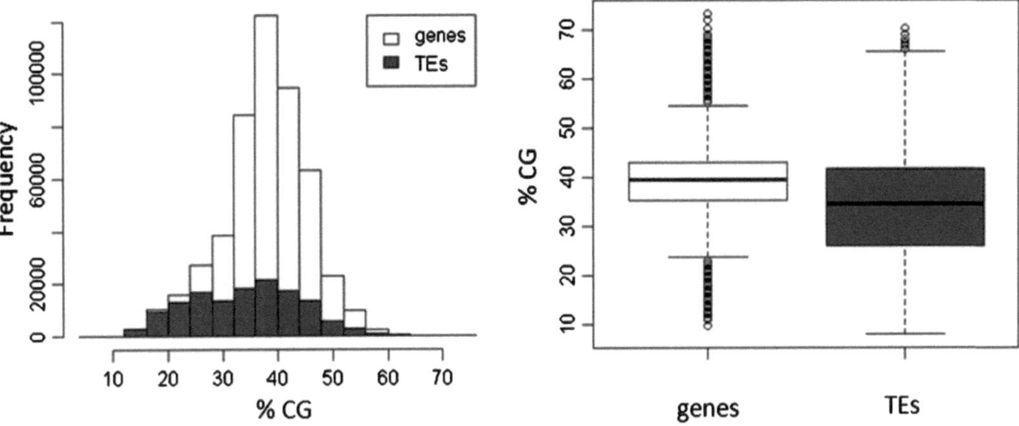

Fig. 12 CG content of genes and transposons. Shown are the CG content distributions of genes and transposable elements (TEs)

to strong dye biases (Fig. 11b). We find a clear negative linear relationship between CG content and signal intensity; that is, signal intensity is highest for probes with low CG content and lowest for probes with high CG content. By contrast, CG content appears to have little influence on the signal intensity in the R/G data (Fig. 11c), and the DS displays intermediate levels of CG bias (Fig. 11d). *See* **Note 2** for recommendations concerning the labeling of the IP and input DNA.

Royce et al. [30] considered normalizing tiling array signals for the CG content of probes. While this procedure may work for some applications, such as transcription factor binding data (ChIP-chip), its application to gene expression tiling arrays has been shown to lead to overnormalization and hence to a loss of signal information [31]. Overnormalization is expected to be even more drastic in MeDIP-chip data where CG content is correlated with DNA methylation levels. Correcting for CG content, in this case, will reduce signal intensities arising from probes with true positive methylation measurements. A simple solution to bypass these issues is to work exclusively with R/G data where CG bias appears to be minimal (Fig. 11c, *see* **Note 2**).

4 Concluding Remark

Although the mapping resolution of MeDIP-chip (~165 bps) is much lower than the single cytosine measurements that can be achieved with WGBS-seq, this array-based technology provides a level of resolution that should be sufficient for the detection of most functionally important differentially methylated regions. MeDIP-chip requires fewer bioinformatic resources and therefore scales more easily to large samples. Provided several experimental and data preparation steps are followed (*see* **Note 1–3**), MeDIP-chip presents a viable alternative to WGBS-seq in future population epigenetic studies.

5 Notes

1. *Recommendations for data preparation*: Prior to MeDIP-chip analysis, potentially cross-hybridizing probes should be removed. They typically show signal intensities similar to probes that correspond to genomic regions with high methylation levels (Fig. 3). Failure to remove cross-hybridizing probes can therefore result in the detection of a large number of false positives. However, removal of these probes will result in loss of measurement coverage; but this drawback is no different from sequencing-based approaches where short reads that do not map uniquely are usually excluded.
2. *Recommendations for dye-labeling*: Dye-related biases can pose serious concerns in dual-channel MeDIP-chip. Labeling the immunoprecipitate (IP) DNA with Cy3 (green) and the control DNA (input) with Cy5 (red) (i.e., G/R data) introduces strong signal distortions that significantly compromise measurement accuracies. These biases are particularly pronounced in genomic regions with low methylation and low CG content. This signal bias disappears when the opposite labeling strategy is employed (labeling IP with Cy5 and input with Cy3, i.e., R/G data). As a result of the G/R dye bias, dye-swap experiments in MeDIP-chip always perform worse than the R/G data alone, despite the fact that DS consists of twice as much data. Hence, despite its routine use in expression micro-array studies, we do not recommend the use of dye-swaps in dual-channel MeDIP-chip. This means that experimental costs can be reduced by a factor of two without loss of measurement information.
3. *Recommendations for data analysis*: The classification of probes as methylated or unmethylated requires a sound statistical approach. The best methods for MeDIP-chip are variants of HMMs [28]. The assumptions of HMMs are fundamentally consistent with the data properties arising from MeDIP-chip experiments. These assumptions are the following: (1) A probe signal is a noisy proxy for an underlying (unobserved) methylation state and (2) methylation states are spatially correlated along the genome owing to the array design and the propensity of DNA methylation to occur in clusters. Our application of a recent HMM designed for *Arabidopsis* MeDIP-chip [15] resulted in a genome-wide false-positive rate of about 0.02 and

false-negative rate of about 0.19 (a combined rate of 0.21) for the R/G data. This relatively high false-negative rate implies that the application of this HMM misses regions with low methylation levels. Less customized methods may yield even higher misclassifications. In population studies, this limitation will restrict the calling of DMRs to clear methylation differences between individuals (e.g., no methylation versus high methylation) and will likely fail to detect more subtle DMRs (e.g., no methylation versus low methylation). One way to improve this situation is to consider at least one additional technical R/G replicate.

Acknowledgements

This work was supported by grants from the Netherlands Organization for Scientific Research (NWO) (to F.J. and M.C.-T) and the Netherlands Bioinformatics Centre (NBIC) (to R.W.). Work in the Colot lab is supported in part by the European Union Network of Excellence EpigeneSys.

References

1. Law JA, Jacobsen SE (2010) Establishing, maintaining and modifying DNA methylation patterns in plants and animals. Nat Rev Genet 11:204–220
2. Laird PW (2010) Principles and challenges of genome-wide DNA methylation analysis. Nat Rev Genet 11:191–203
3. Zhang X, Yazaki J, Sundaresan A, Cokus S, Chan SW, Chen H (2006) Genome-wide high-resolution mapping and functional analysis of DNA methylation in *Arabidopsis*. Cell 126:1189–1201
4. Zilberman D, Gehring M, Tran RK et al (2007) Genome-wide analysis of *Arabidopsis thaliana* DNA methylation uncovers an interdependence between methylation and transcription. Nat Genet 39:61–69
5. Lister R, O'Malley RC, Tonti-Filippini J, Gregory BD, Berry CC, Millar AH, Ecker JR (2008) Highly integrated single-base resolution maps of the epigenome in *Arabidopsis*. Cell 133:523–536
6. Cokus SJ, Feng S, Zhang X, Chen Z, Merriman B, Haudenschild CD et al (2008) Shotgun bisulphite sequencing of the *Arabidopsis* genome reveals DNA methylation patterning. Nature 452:215–219
7. Schmitz RJ, Schultz MD, Lewsey MG et al (2011) Transgenerational epigenetic instability is a source of novel methylation variants. Science 334:369–373
8. Becker C, Hagmann J, Müller J et al (2011) Spontaneous epigenetic variation in the *Arabidopsis thaliana* methylome. Nature 480:245–249
9. Boyes J, Bird A (1992) Repression of genes by DNA methylation depends on CpG density and promoter strength: evidence for involvement of a methyl-CpG binding protein. EMBO J 11:327–333
10. Lorincz MC, Schübeler D, Hutchinson SR, Dickerson DR, Groudine M (2002) DNA methylation density influences the stability of an epigenetic imprint and Dnmt3a/b-independent de novo methylation. Mol Cell Biol 22:7572–7580
11. Johannes F, Porcher E, Teixeira F, Saliba-Colombani V, Simon M, Agier N et al (2009) Assessing the impact of transgenerational epigenetic variation on complex traits. PLoS Genet 5:e1000530
12. Colomé-Tatché M, Cortijo S, Wardenaar R, Morgado L, Lahouze B, Sarazin A et al (2012) Features of the *Arabidopsis* recombination landscape resulting from the combined loss of sequence variation and DNA methylation. Proc Natl Acad Sci USA 109:16240–16245
13. Weber M, Davies JJ, Wittig D, Oakeley EJ, Haase M, Lam WL, Schübeler D (2005) Chromosome-wide and promoter-specific analyses identify sites of differential DNA

methylation in normal and transformed human cells. Nat Genet 37:853–862
14. Down TA, Rakyan VK, Turner DJ, Flicek P, Li H, Kulesha E et al (2008) A Bayesian deconvolution strategy for immunoprecipitation-based DNA methylome analysis. Nat Biotechnol 26:779–785
15. Cortijo S, Wardenaar R, Colomé-Tatché M, Johannes F, Roudier F, Colot V (2012) Genome-wide analysis of DNA methylation in *Arabidopsis* using MeDIP-chip. Methods Mol Biol 15:2930–2939
16. Frommer M, McDonald LE, Millar DS, Collis CM, Watt F, Grigg GW et al (1992) A genomic sequencing protocol that yields a positive display of 5-methylcytosine residues in individual DNA strands. Proc Natl Acad Sci USA 89:1827–1831
17. Clark SJ, Harrison J, Paul CL, Frommer M (1994) High sensitivity mapping of methylated cytosines. Nucleic Acids Res 22:2990–2997
18. Lister R, Ecker JR (2009) Finding the fifth base: genome-wide sequencing of cytosine methylation. Genome Res 19:959–966
19. Krueger F, Kreck B, Franke A, Andrews SR (2012) DNA methylome analysis using short bisulfite sequencing data. Nat Methods 9:145–151
20. Chen PY, Cokus SJ, Pellegrini M (2010) BS Seeker: precise mapping for bisulfite sequencing. BMC Bioinformatics 11:203
21. Bolstad BM, Irizarry RA, Astrand M, Speed TP (2003) A comparison of normalization methods for high density oligonucleotide array data based on variance and bias. Bioinformatics 19:185–193
22. R Development Core Team (2012) R: a language and environment for statistical computing. R Foundation for Statistical Computing, Vienna, ISBN: 3-900051-07-0. http://www.R-project.org
23. Dobbin KK, Kawasaki ES, Petersen DW, Simon RM (2005) Characterizing dye bias in microarray experiments. Bioinformatics 21:2430–2437
24. Dombkowski AA, Thibodeau BJ, Starcevic SL, Novak RF (2004) Gene-specific dye bias in microarray reference designs. FEBS Lett 560:120–124
25. Martin-Magniette ML, Mary-Huard T, Bérard C, Robin S (2008) ChIPmix: mixture model of regressions for two-color ChIP-chip analysis. Bioinformatics 24:i181–i186
26. Andrews S (2007) ChIPmonk: software for viewing and analysing ChIP-on-chip data. BMC Syst Biol 1(Suppl 1):P80
27. Johannes F, Wardenaar R, Colomé-Tatché M, Mousson F, de Graaf P, Mokry M et al (2010) Comparing genome-wide chromatin profiles using ChIP-chip or ChIP-seq. Bioinformatics 26:1000–1006
28. Seifert M, Cortijo S, Colomé-Tatché M, Johannes F, Roudier F, Colot V (2012) MeDIP-HMM: genome-wide identification of distinct DNA methylation states from high-density tiling arrays. Bioinformatics 28:2930–2939
29. Li W, Meyer CA, Liu XS (2005) A hidden Markov model for analyzing ChIP-chip experiments on genome tiling arrays and its application to p53 binding sequences. Bioinformatics 21(Suppl 1):i274–i282
30. Royce TE, Rozowsky JS, Gerstein MB (2007) Assessing the need for sequence-based normalization in tiling microarray experiments. Bioinformatics 23:988–997
31. Gilbert D, Rechtsteiner A (2009) Comments on sequence normalization of tiling array expression. Bioinformatics 25:2171–2173

Chapter 14

Mapping Genomic Features of Tiling Microarray Data by TileMapper

Hoi-Hung Cheung, Janek Claus, Sumeeta Singh, Chandan Sastry, Owen M. Rennert, Wai-Yee Chan, and Tin-Lap Lee

Abstract

The recent revolution of genomics techniques has allowed the detection of various sequence features and biological variations on whole-genome scale. However, these high-resolution data present significant challenges for experimental biologists to understand and analyze. The conventional way is to use genome browsers to locate and visualize regions of interest. But it lacks user-friendly data mining functionality. Here we present a protocol that allows rapid annotation of genomic coordinate data by using TileMapper. Interesting biological annotations from large-scale genomic data, such as transcriptome analysis, chromatin immunoprecipitation on chip, or methylated DNA immunoprecipitation (MeDIP) studies generated from the tiling microarrays and other platforms, could be analyzed without requiring computational skills. The outputs are saved in tabulated format, which permit flexible and simple processing in spreadsheet software, or to be exported to other pipelines for subsequent analysis.

Key words Annotations, Genomics, ChIP-chip, MeDIP, Transcriptome

1 Introduction

With the completion of various genomes, the sequence complexity suggested that they are more complex than previously envisioned. Traditional microarray platforms are gene centric and are biased to the 3′ untranslated regions. In contrast, tiling microarrays have probes that cover the entire genome or contigs of the genome in an unbiased fashion with higher resolution independent of experimentally or computationally derived evidence [1, 2]. Its applications cover a wide range of genomic analyses, including whole-genome transcriptome analysis, detection of transcription factor-binding sites, chromatin modifications, and epigenetic modifications such as methylated DNA immunoprecipitation (MeDIP) by chromatin immunoprecipitation on chip (ChIP-chip) approach. However, the implementation of such high-resolution data in the global analysis of large mammalian genomes presents significant challenges for

Tin-Lap Lee and Alfred Chun Shui Luk (eds.), *Tiling Arrays: Methods and Protocols*, Methods in Molecular Biology, vol. 1067,
DOI 10.1007/978-1-62703-607-8_14, © Springer Science+Business Media New York 2013

experimental biologists. For instance, the Integrated Genome Browser (IGB) from Affymetrix provides an excellent platform for browsing the raw signal or transcribed fragments (Transfrags) from the tiling data with intuitive graphical interfaces. It allows signal filtering and aligning with various genomic annotations. However, it lacks user-friendly data mining functionality to identify the tiling signals that match particular genomic annotations for subsequent data manipulation. Analyses like discovering novel transcripts based on non-Refseq-associated ESTs in transcriptome experiment or identifying transfrags that map to the binding sites in ChIP-chip or CpG islands in MeDIP experiment are not possible. To proceed to the data mining level, users need to have some programming skills and understand the underlying database schema for data access.

TileMapper is an online application that annotates the transfrags from tiling microarray data by mapping to various genomic annotations in tiling array analyses. TileMapper provides a user-friendly environment and flexible options on annotation selection. The result file can be analyzed by spreadsheet application or further processed in downstream pipelines to relate the data and perform interaction analysis with other datasets, which will be useful for gaining better understanding of the regulatory networks and allow novel gene discovery and can provide other insights toward genome complexity.

2 Materials

2.1 Computer Hardware and Data Source

1. TileMapper is a Web-based tool. It requires a computer or a mobile device with Internet connection (*see* **Note 1**).

2.2 Software

1. Affymetrix GeneChip Command Console (AGCC) Software, or GeneChip Operating Software (GCOS) version 1.3 or higher.
2. Affymetrix Tiling Analysis Software (TAS).
3. IGB from Web start (http://www.affymetrix.com/partners_programs/programs/developer/tools/download_igb.affx) or through local installation.
4. The software are available for free for Affymetrix users, downloadable at http://www.affymetrix.com.
5. TileMapper provides a simple interface for users to identify important genomic annotation features like Refseq, EST, mRNA, promoter, and CpG island. It is computer platform independent and supports popular Web browsers including Mozilla Firefox, Microsoft Internet Explorer, Safari, and Google Chrome. TileMapper adopts BED file format for input (*see* **Note 2**).

3 Methods

In this protocol, we first show how the BED files can be obtained from popular genomic platform, such as Affymetrix IGB or from public resources including Gene Expression Omnibus (GEO) at NCBI and genomic track data from UCSC genome browser. Then we illustrate how to map the selected annotations. Finally we explain the result files and how to get imported in spreadsheet software.

The front page of TileMapper is divided into three sections: the BED file input box on the top panel, the annotation menu in the middle, and the e-mail notification box at the bottom. Generally, the annotation analysis can be finished in four simple steps: (1) select the BED file, (2) select the annotation, (3) send the request to TileMapper, and (4) browse/download results. The overview of TileMapper workflow is illustrated in Fig. 1.

3.1 Generating BED Files from Affymetrix Integrated Genome Browser

1. The tiling microarray data generated from Affymetrix TAS or software from other platforms could be loaded into Affymetrix IGB for data visualization first.
2. IGB is an application intended for visualization and exploration of genomes and corresponding annotations from multiple data

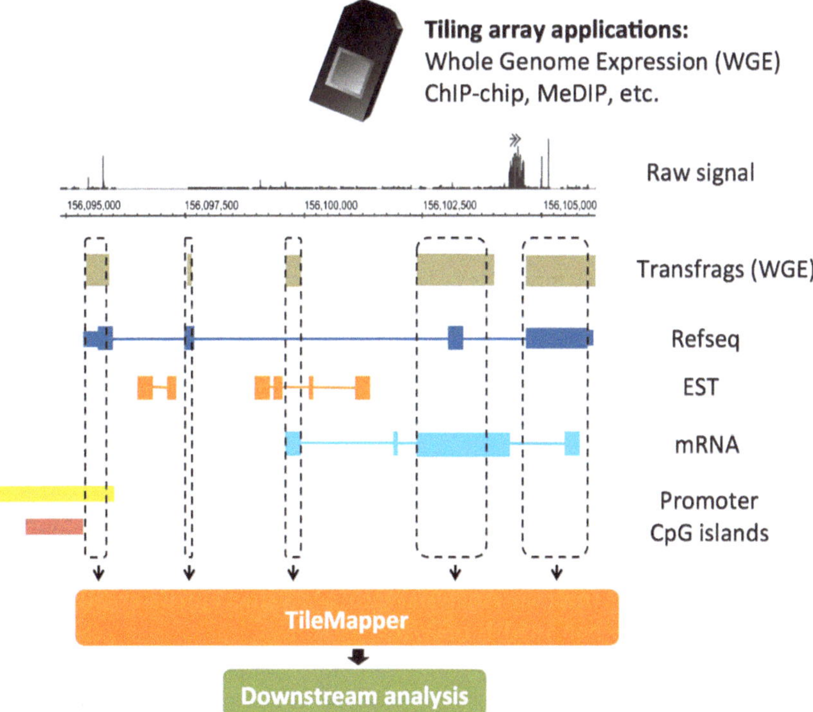

Fig. 1 Overview of TileMapper. TileMapper provides a simple interface for users to map important genomic annotation features to the dataset from various tiling microarray experiments. Significant raw signals from tiling microarray experiments, GEO, or UCSC database in the form of BED file format can be applied. The mapped file can be used for downstream analysis

sources. Launch the program at http://www.affymetrix.com/partners_programs/programs/developer/tools/download_igb.affx. Alternatively, it can be downloaded from http://genoviz.sourceforge.net. IGB is Java dependent and therefore Java has to be installed before running the program. Java runtime can be found at http://www.java.com (*see* **Note 3**).

3. Genomic coordinate data (either derived from *p*-values or signal files in BAR format) generated from Affymetrix TAS (http://www.affymetrix.com/partners_programs/programs/developer/TilingArrayTools/index.affx) or from other analytical platforms are imported to the IGB by selecting the "Open" function under "File" (*see* **Note 4**).
4. To create a BED file that represents the dataset from TAS, right-click on the track annotation region to bring up the "save annotations" option. Select and click "save tier as BED file" to create the file. The BED file will contain the genomic coordinates for the start and end position for each feature (*see* **Note 5**).

3.2 Downloading Genomics Coordinate Data from NCBI GEO Database

1. In addition to in-house experimental data, BED files from other published experiments can be downloaded from public database such as NCBI GEO. For example, we have applied Affymetrix Human Tiling 2.0R Set to perform MeDIP data on testicular germ cell tumor (TGCT), where 26 % of differentially methylated genes showed differential gene expression [3].
2. Download the methylation data from NCBI GEO database under series ID GSE15220. Genomic DNA was extracted from the cultured TGCT cell line Ntera2 and a normal testis cell line (CRL-7002). Methylation-enriched DNA was amplified and hybridized to Human Tiling Array 2.0R chips (Affymetrix). Triplicate sets of hybridization were performed from three independent MeDIP experiments for each cell line.
3. Analyze the raw data by TAS by using quantile normalization method. Arrays from each group (cancer vs. normal) were quantile normalized and differential methylation between groups of tumor and normal was compared by choosing the "two-sample comparison analysis" option in TAS.
4. Peform two-sided test to evaluate both hypermethylation and hypomethylation. Transfrags (or DMRs) were generated by Interval Analysis with a *p*-value cutoff at 20 ($p<0.01$) and a bandwidth of 275.
5. Transfrags generated by *p*-value cutoff with a positive signal difference were defined as hypermethylation while those of negative difference were defined as hypomethylation.
6. Save the files to the computer. A total of 22,452 hypermethylated (GSE15220_Chip_ALL_HYPER.bed) and 12,756

hypomethylated (GSE15220_Chip_ALL_HYPO.bed) regions will be revealed.

7. Alternatively, download the BED files that represent the hypermethylated and hypomethylated track; it can be directly obtained at the following URL: http://www.ncbi.nlm.nih.gov/geo/query/acc.cgi?acc=GSE15220.

3.3 Downloading Genomics Coordinate Data from UCSC Genome Browser

1. Various tiling microarray experiment data are also publicly available in UCSC server. For example, data of the ENCyclopedia Of DNA Elements (ENCODE) project that contains 1 % of functional elements in the human genome sequence can be retrieved (http://genome.ucsc.edu/ENCODE/).
2. The expression data from the tiling array platforms in ENCODE could be download and exported. For example, the data generated in transcriptome study on HL-60 cells using Affymetrix ENCODE tiling arrays is located at http://genome.csdb.cn/cgi-bin/hgTables?db=hg18&hgta_group=encodeTxLevels&hgta_track=encodeAffyRnaTransfrags&hgta_table=encodeAffyRnaHl60SitesHr32&hgta_doSchema=describe+table+schema [4, 5] (*see* **Note 6**).
3. The whole dataset can be exported in the form of data track and saved in BED format for TileMapper analysis.

4 Data Analysis

To identify the annotations associated with the input genomic coordinates, click "browse" to load the BED file, and then select the genome followed by chromosome and annotation feature. Currently, common genome annotations such as Refseq, EST, mRNA, promoter, and CpG island are included.

4.1 Upload the BED File

1. Type http://tilemapper.nichd.nih.gov/ in the URL box of Web browser, and press ENTER.
2. The TileMapper tool will show up. Identify the BED file input box.
3. Upload the BED file from either one of the approaches in Subheading 3.
4. Click "Browse…" next to the BED file box to upload the BED file.
5. Click on "Use sample bed file" if you would like to test TileMapper. A sample BED file is included. Click "Show" to download the test BED file. View the file in a text editor or in Microsoft Excel.

4.2 Select Annotations to Be Mapped

1. To select annotations to be mapped, first go to the "Genome" selection box to select the desired species to be mapped. TileMapper currently supports human and mouse genome annotation mapping.
2. Next, select the chromosome number to be annotated.
3. Select the desired annotation to be mapped by checking the "Annotated features" box. The box contains five popular genome annotations, including Refseq, mRNA, ESTs, Promoter, CpG islands, and All. The Refseq, mRNA, and EST annotations are useful for mapping transcripts from transcriptome study, while Promoter and CpG are useful in ChIP analysis such as MeDIP. To retrieve all five annotations, select "All" option in the box.
4. Fill in the e-mail field if the user wants to get process completion notification e-mail. To start the analysis, click "Start mapping Process" to begin the mapping. A notification e-mail will be sent to confirm the data submission (if e-mail entered). When processing is completed, an e-mail will be sent with the result file attached. Otherwise the results can be directly downloaded from the TileMapper Webpage.
5. For example, to identify genes affected by aberrant DNA methylation in TGCT study mentioned in Subheading 3.2, simply select Refseq annotation. Complete information related to affected exon/intron and associated gene ID will be analyzed. To know more about the transcriptional information about the DNA methylation, CpG island and promoter can be also selected and analyzed in a similar way.
6. To experiment the whole workflow, the user may use the example dataset provided. Select "Use sample bed file" under BED file input field to analyze the example dataset. Click "Show" to show the contents in the data file example (*see* **Notes 7** and **8**).

4.3 Viewing the Output File

The output of TileMapper is in tab-delimited text format (.txt) and can be visualized in spreadsheet software such as Microsoft Excel.

1. To export in Microsoft Excel, select "Open" under "File." Identify the file as tab-delimited in the pop-up window and click "finish" to complete the import procedure.
2. Below is the partial mapped result for the sample transfrags to mouse Refseq chromosome 1 annotation:

```
   Criteria:Genome:mouse_35;Chromosome:chr1;
Annotation:refseq
   Chromosome  Start End Size  ID  Region
   chr1 3044450 3044553 103 NM_022114,NM_199454,
In1,In1,
   chr1 3050795 3050869 74 NM_022114,NM_199454,
In1,In1,
```

```
chr1 3051033 3051799 766 NM_022114,NM_199454, In1,In1,
chr1 3143817 3143885 68 NM_022114,NM_199454, In2,In2,
chr1 3144052 3144131 79 NM_022114,NM_199454, In2,In2,
```

The first four columns are the information related to the transfrags, starting with chromosome number, start position, end position, and the size of transfrag. The fifth column is Refseq id(s) mapped to the transfrag and to which region it maps to, in terms of exon (Ex) or intron (In) followed by the Ex/In number (sixth column). If more than one id is matched, the ids will be separated by comma. For example:

```
chr1 3649724 3649841 117 NM_001126240, NM_005427 In2,In4
```

This indicates that the given genomic feature maps to intron 2 of NM_001126240 and intron 4 of NM_005427.

3. If more than one annotation is selected in the query, the annotations will be listed in order of Refseq, mRNA, ESTs, Promoter, and CpG islands.

5 Notes

1. The Internet connection speed affects uploading and downloading times for input and output datasets. A PC with at least 1 Gb of memory is recommended when working with large dataset.
2. TileMapper annotates large-scale genomic data in Browser Extensible Display (BED) format. BED file is a flexible file format that contains the genomic coordinate information on significant regions. It is a common file format generated from high-throughput genomic platforms like tiling microarrays.
3. IGB can be started from Web or through local installation. To run IGB from Web, make sure that Java Web Start (www.java.com) is installed. IGB requires java at least version 1.4.2. However, version 1.6 is recommended and any more recent version can also be used. Java Web Start is included with recent versions of the Java 2 Runtime Environment (JRE) and Java 2 Standard Edition SDK (J2SE SDK). The minimum memory requirement for IGB alone is 256 mb of system memory; therefore a system of at least 512 mb is required. However, we recommend to run IGB in a system with at least 2 gb of system memory. To load the intensity data from all the chips in triplicate setting, start IGB with 1.5 gb memory requirement. Use of lower memory requirement options will cause slow down or program crash.

4. Other than known gene/position search, IGB also offers pattern search functionality. Regular expression pattern search is very useful for looking for specific sequence features or finding instances of a sequence containing unknowns. A complete list of regular expression syntax can be found at http://java.sun.com/j2se/1.5.0/docs/api/java/util/regex/Pattern.html.
5. In addition to Refseq annotation, IGB also provides access to other annotation information, which can be retrieved from NetAffx or DAS/2 server.
6. This track shows the location of sites showing transcription for all ENCODE regions in several cell types, using Affymetrix arrays. Retinoic acid-stimulated HL-60 cells were harvested after 0, 2, 8, and 32 h. Purified cytosolic polyA+ RNA from unstimulated GM06990 and HeLa cells, as well as purified polyA+ RNA from the RA-stimulated HL-60 samples, was hybridized to Affymetrix ENCODE oligonucleotide tiling arrays, which have 25-mer probes tiled every 22 bp on average in the non-repetitive ENCODE regions. Clustered sites are shown in separate subtracks for each cell type and for each of the four timepoints for RA-stimulated HL-60.
7. TileMapper presents a simple interface for biologists to identify large-scale genomic data. Importantly, the tool does not require any computation skills to work with. Other online tools like Galaxy could do similar annotation work, but it does not provide any information on exon/intron analysis. Furthermore, one has to spend some time to learn how to operate it. Without involving too many technical terms and operations, TileMapper provides a much user-friendly way to get the task finished within three simple mouse clicks. Although it was developed primarily for tiling microarray analysis, it can also be applied to other large-scale datasets like RNA-seq or ChIP-seq.
8. The mapping process usually takes less than 1 min to finish; in some cases, especially when selecting all annotations, it may take up to 10 min to complete. The fast processing of the dataset attributed to the optimized Java code implemented in a layered architecture. In our preliminary analysis, the code is up to three times faster than the Perl-based code. The code is freely available upon request. We intentionally disabled the full chromosome analysis feature to prevent from overloading of our server. We will release this function when our server cluster gets a complete hardware upgrade. TileMapper has been tested with various tiling data from our in-house Affymetrix tiling 1.0R/2.0R experiments and ENCODE data from other tiling platforms at UCSC server since 2008. We will include more annotation datasets in the next version.

Acknowledgments

This study was supported by the Intramural Research Program of the Eunice Kennedy Shriver National Institute of Child Health and Human Development, National Institutes of Health and General Research Fund from Research Grant Council, Hong Kong.

References

1. Gregory BD, Yazaki J, Ecker JR (2008) Utilizing tiling microarrays for whole-genome analysis in plants. Plant J 53:636–644
2. Mockler TC, Chan S, Sundaresan A, Chen H, Jacobsen SE, Ecker JR (2005) Applications of DNA tiling arrays for whole-genome analysis. Genomics 85:1–15
3. Cheung HH, Lee TL, Davis AJ, Taft DH, Rennert OM, Chan WY (2010) Genome-wide DNA methylation profiling reveals novel epigenetically regulated genes and non-coding RNAs in human testicular cancer. Br J Cancer 102:419–427
4. Cawley S, Bekiranov S, Ng HH, Kapranov P, Sekinger EA, Kampa D et al (2004) Unbiased mapping of transcription factor binding sites along human chromosomes 21 and 22 points to widespread regulation of noncoding RNAs. Cell 116:499–509
5. Kapranov P, Cawley SE, Drenkow J, Bekiranov S, Strausberg RL, Fodor SP, Gingeras TR (2002) Large-scale transcriptional activity in chromosomes 21 and 22. Science 296: 916–919

Acknowledgments

Index

Tin-Lap Lee and Alfred Chun Shui Luk (eds.), *Tiling Arrays: Methods and Protocols*, Methods in Molecular Biology, vol. 1067, DOI 10.1007/978-1-62703-607-8, © Springer Science+Business Media New York 2013

MIX
Papier aus verantwortungsvollen Quellen
Paper from responsible sources
FSC® C105338

If you have any concerns about our products,
you can contact us on
ProductSafety@springernature.com

In case Publisher is established outside the EU,
the EU authorized representative is:
Springer Nature Customer Service Center GmbH
Europaplatz 3, 69115 Heidelberg, Germany

Printed by Libri Plureos GmbH
in Hamburg, Germany